Understanding Cell Structure

Understanding Cell Structure

MARTIN W. STEER
Reader in Botany at The Queen's University of Belfast

CAMBRIDGE UNIVERSITY PRESS

CAMBRIDGE
LONDON NEW YORK NEW ROCHELLE
MELBOURNE SYDNEY

Published by the Press Syndicate of the University of Cambridge
The Pitt Building, Trumpington Street, Cambridge CB2 1RP
32 East 57th Street, New York, NY 10022, USA
296 Beaconsfield Parade, Middle Park, Melbourne 3206, Australia

© Cambridge University Press 1981

First published 1981

Typeset by Asco Trade Typesetting Limited, Hong Kong
Printed in Great Britain at the University Press, Cambridge

British Library Cataloguing in Publication Data

Steer, Martin William
Understanding cell structure.

1. Cells
I. Title
574.87   QH581.2   80-41202
ISBN 0 521 23745 9 hard covers
ISBN 0 521 28198 9 paperback

Contents

Preface

The first serious biological microscopists, Robert Hooke and Antonie van Leeuwenhoek, were fascinated by the fine scale of the structures they observed. Both estimated the numbers of structures present in a unit volume (usually a cubic inch) and were surprised at the magnitude of the results, to the extent that van Leeuwenhoek felt compelled to obtain affidavits from independent witnesses testifying to the accuracy of his claims.

These revelations of light microscopy, remarkable as they are, do little to prepare the observer for the astonishing wealth of fine detail revealed by the electron microscope. There is no doubt that the architecture of a cell is impressive, and many studies have helped to document the structure of a great range of cell types. Even more remarkable than the structure is the functioning of cells, and microscopists have struggled to place dynamic and functional interpretations on the static images from electron microscopes. The task is made more complex by the necessity to transform two-dimensional information from micrographs back to the three-dimensional structure of cells. Few books are available to instruct the beginner in the interpretation of these images, compared with the wealth of material available on both the previous steps, the preparation of specimens for microscopy and the use of microscopes, and on the final conclusions, the structure and function of cell components. This book attempts to fill the gap by providing a guide to the main considerations in the understanding of cell structure, although it is not a substitute for actual experience.

A microscopist must have the ability to mentally reconstruct images of the whole cell from micrograph snapshots, and turn these around in the mind and observe them from all angles. In other words, an ability to climb down inside a cell and wander about, examining the architecture and looking at the possible ways in which it might function. This ability to handle visual images is not always accompanied by an ability to handle abstract mathematical relationships. Hence the section on quantitative microscopy, which makes up most of the book, has been written, as far as possible, in a non-mathematical form, in an attempt to persuade all microscopists that this approach is not beyond them and would be of considerable value in their studies.

My own interest in this subject was stimulated by Brian Gunning (now Professor of the Department of Developmental Biology, Research School of Biological Sciences, ANU) who first included it in his teaching to the students of the Cell Biology course here about ten years ago. His schedule for this part of the course has found its way into many laboratories around the world, and has formed the basis for many ideas in this book. Acknowledgement must also be made to the students of this course who, over the intervening years, have assailed me with a wealth of serious questions. Searching out answers to these questions has improved my own understanding of this subject and increased my awareness of the problems faced by beginners in this field. I have tried to incorporate this experience into the text. These students have one major advantage over me, their previous excellent instruction in the field of statistical analysis. For advice and criticism of this section of the book I am indebted to Reg Parker; the remaining errors in this section are, of course, mine. My thanks are due to R. Reed of the Queen's University Electron Microscope Unit for his assistance and co-operation with the use of the Cambridge S-180 STEM unit and the Philips 400 microscope. Especial thanks are due to George McCartney for running the Botany Electron Microscope Unit, and for printing the figures for this book. The Cell Biology research students are to be thanked for their encouragement and patience while I was writing this book. My wife, Lesley, not only provided encouragement, but also practical assistance: the meticulous typing of the final manuscript; for both I am very grateful.

MWS
Belfast, May 1980

PART I

Cells and microscopes

Introduction

Microscopy provides a unique insight into the remarkable structural organisation of cells. Yet if we are to improve our understanding of cell structure we need to do more than just look at a micrograph and describe the components present. Even before we start, we need to be aware of the processes involved in transforming part of a living cell to a microscope image; we also need to know something of the general properties and functions of the cell and of the components in it. This background knowledge will be largely assumed in this book; there are many texts dealing with the processing of cells and tissues for microscopy and the principles and operation of light and electron microscopes (for example, Hayat, 1970, 1972, 1973, 1974; Glauert, 1974a, b, 1978; Grimstone, 1977; Hall, 1978). Similarly the structure and function of cells are described in many publications (for example, De Robertis, Nowinski and Saez, 1970; Wolfe, 1972; Hall, Flowers and Roberts, 1974; Hopkins, 1978; Pickett-Heaps, 1975; Gunning and Steer, 1975a; Novikoff and Holtzmann, 1976). Here we will be concerned with methods and techniques for the interpretation of micrographs, only introducing and emphasising those parts of the background information that are necessary for an understanding of this approach.

Cell structure

We can analyse cells and make a list of all the chemicals present with a fair amount of precision. Yet adding these chemicals together in the correct proportions and concentrations does not reconstitute a new cell; clearly something is missing. The missing ingredient is organisation, the organisation of the components to form an integrated, dynamic system. Living cells are a delicate balance of physical and chemical interactions between the component molecules.

The composition of living systems is in stark contrast to the non-living components of the Earth's surface. The latter consist of a relatively small number of molecular species present in very large amounts. They are sometimes highly ordered, forming regular crystalline arrays, and even many apparently heterogeneous materials are composed of mixtures of particles, each having a uniform composition. This is quite unlike living systems which are relatively sparse in occurrence, but consist of very diverse mixtures of molecular species contained within very small units, the cells.

How have these 'living' collections of molecules arisen? It is not the purpose of this book to discuss the numerous conjectures about the origin of life. We can, however, take up these proposals at a stage common to all of them: the formation of a collection of chemicals, isolated by a membrane barrier from the environment, capable of taking in simple (low molecular weight) molecules and joining them together to form complex (high molecular weight) molecules by catalytic (enzymic) reactions.

It is possible that present-day cells have evolved from an ancestor with these characteristics. Certainly they have evolved from a common ancestor, judging by the ubiquity of the genetic code and systems for converting this information from a linear form into three-dimensional catalysts, the enzymes. For an individual cell living today, the route from past to present has been long and tortuous, beset by innumerable changes of environment and competition from other cells.

What were, and are, the evolutionary forces determining the pattern of cellular organisation? They are the physical processes whereby substrates arrive at enzyme sites on protein catalysts and the chemical reactions that occur when they arrive. Increased versatility of the catalysts within a cell is advantageous, but leads to conflicting requirements for local environments (pH, redox, etc.) by the different enzymes. Internal compartmentation provides the opportunity for specialised local environments within the cell volume and additional sites for enzymes on and within the bounding compartment membranes.

As far as we can judge all the basic cell structures and systems were evolved by unicellular organisms. Multicellular states are a relatively recent innovation and the constituent cells have far more in common with each other than might be guessed from examining the diverse morphologies of plants and animals.

This common ancestral origin is of great importance. It is one of the foundation stones of cell biology, enabling information and understanding gleaned from one source to be applied to a whole range of living systems.

Microscopy and microscope images

Two basic approaches are used to investigate the structure and function of cells. One may be called the biochemical approach. In general it involves breaking open the cells to release the components and fractionating them to give preparations of each type of component. These can be analysed for their gross chemical composition and tested for enzyme activity. Eventually this will lead to an understanding of the total activity within each component and the way in which they interact in the whole cell. The second approach is to examine cell structure using a microscope to determine the morphological characteristics of each component and its spatial relationship to the rest of the cell.

Each of the approaches has its drawbacks, the biochemical approach destroys the complex three-dimensional organisation of the cell which is so elegantly portrayed by microscopy. On the other hand microscopy is not able to provide the specific information about the composition of the cell components and their enzyme activities that is obtainable by biochemical methods. In practice both methods are employed, and may be combined, for example in the microscopy of cell fractions from tissue homogenates and the localisation of enzymes in the cell structure by cytochemical methods.

There is a further general difference in these two approaches to the study of cell structure. Biochemical analyses yield experimental data in numerical form which can be analysed objectively. However, microscopy yields visual images which are not easy to analyse objectively, and so are open to subjective interpretation.

In order to understand these images we first need to understand how they are formed in the microscope. The use of a microscope system to analyse cell structure is limited by two constraints. One is the resolving power of the instrument, that is, the smallest space between two particles that can be detected, and the other is the contrast in the system, that is, the ability of the system to discriminate between components of different chemical composition.

Light microscopes are limited by the wavelength of light to a resolution limit of about 200 nm. Electron microscopes are limited by the design of the electromagnetic lenses to a resolution of about 0·2 nm.

Light microscope systems depend on two types of contrast to distinguish different parts of the specimen. One depends on the selective absorption of specific wavelengths of light, from the broad range present in a white light source, giving a colour to that particular component in the image. Alternatively, the energy from the absorbed wavelengths, usually short, may be re-emitted at a longer wavelength (fluorescence) which can be detected with the aid of suitable filters. The other is dependent on the density of the specimen, which slows down the wavefront of light passing through it; this can be converted into an intensity difference at the image plane by a phase contrast illumination system.

In the electron microscope, the image depends on differences in the number of electrons arriving at different parts of the image plane. This numerical density can be affected (lowered) by deflecting, or scattering, electrons out of the beam, so giving darker patches on the image plane. Elements of high atomic number in the specimen are more efficient scatterers of electrons than those with low atomic numbers.

Most cell components are very similar to one another in their physical properties and basic chemical composition. They all tend to interact in a similar way with either light or electron radiation in microscope systems. Relatively few molecules are coloured, while the intensity differences in a phase system are only characteristic of the gross features of the components, not the individual structural differences. Most molecules of biological systems are composed of elements of low atomic number which interact in similar ways with the beam in an electron microscope. These deficiencies in contrast are partly overcome by the use of stains, that is, additional molecules introduced from outside that bind to certain more or less specific sites in the cell imparting enhanced contrast to them.

Coloured dyes are employed as stains in light microscopy while salts of heavy metals (principally osmium, uranium and lead) are used in electron microscopy. Unfortunately the variety of specific stains available does not match the enormous diversity of components present in a living cell, although the use of labelled (stained) antibodies and lectins as intermediate locators of specific molecules has considerably expanded the range of components detectable.

Both transmission light and electron microscopes depend on radiation passing through the specimen to form an image. Tissues and cells are usually opaque to such radiations, a problem that can be overcome by slicing, or sectioning, them to produce a thinner specimen. This constraint has led to the development of a set of methods and techniques designed to preserve (fix) the cell components as effectively as possible while they are infiltrated and embedded in a hard block, mechanically strong enough to support them during the sectioning process. Staining the specimens may be carried out before sectioning, but is usually performed afterwards.

Images in both light and electron microscopes are recorded on sheets of photographic film, which are processed and printed on photographic paper. Both the film and the final micrographs portray two-dimensional images of the three-dimensional sections. Each part of the final image consists of the superimposition of all the images from that part of the specimen. If the sections are sufficiently thin compared with the dimensions of the components sectioned they may be regarded as two-dimensional samples of the original three-dimensional cells.

Image interpretation

Microscope images of sectioned cells reveal the range of structures present and the complexity of their organisation. We have seen that it is possible to take advantage of the resolution offered by microscope systems only if we subject the cells to extreme chemical and physical treatments, and these destroy the living properties of the system. Somehow we have to use these images to provide information about the delicately balanced system from which they are derived.

Merely observing and describing the cell structures present in the images is insufficient. It is necessary to determine the composition and activities, or functions, of them so that their contribution to the living cell may be assessed. At first sight this task appears formidable, however, the common evolutionary origins of cells enables relationships between structure and function in one cell type to be extrapolated to other cells with a reasonable degree of confidence. This task is made somewhat easier by the observation that specialised cells contain a greater than usual proportion of the components whose functions are directly involved in these specialised activities. Studies of this general type have established the basic roles of most cell structures, so that it is now possible to predict the general functions of a cell by a qualitative examination of microscope images from its cytoplasm. This qualitative approach is discussed in Part II.

Detailed studies of cell structure are concerned with the control of development and activity, the effects of drugs and toxic substances and of diseases. These comparative studies are concerned with the absolute and relative levels of components in cells. Here quantitative information is required so that objective assessments can be made of the structure under different conditions.

Fortunately the problems involved in obtaining information about intact cells from two-dimensional images are similar to those encountered in materials science, in the examination of rocks, minerals, alloys, etc. A number of principles have been established relating the information available from a two-dimensional sample of the specimen (a polished surface) to the three-dimensional structure of the whole (e.g. Underwood, 1970). These were successfully adapted and applied to microscope images of cells in the 1960s and have been reviewed and extended over the past ten years (Weibel, 1969; Weibel and Bolender, 1973; Solari, 1973, 1975; Briarty, 1975; Rohr, Oberholzer, Bartsch and Keller, 1976; Williams, 1977; James, 1977; Bolender, 1978; Weibel, 1979). This quantitative approach to the interpretation of micrographs is discussed in Part III, with examples and a discussion of some applications in Part IV.

PART II

Qualitative Analysis

Visual interpretation

An electron micrograph can be a bewildering array of lines, dots and fine detail. Visual interpretation of these images requires a background knowledge of the general appearance of cell components. This should be acquired by consulting the widest possible range of previous publications, starting with atlases of cell structure (e.g. Fawcett, 1966; Ledbetter and Porter, 1970; Porter and Bonneville, 1973; Gunning and Steer, 1975b).

An essential prerequisite for the interpretation of electron micrographs is a thorough understanding of the cells and tissues at the light microscope level. When this is established and the first electron micrographs are taken it will be necessary to identify the cell components. The following paragraphs provide an outline description of the basic cell components, the texts referred to above should be consulted for further information.

The first task is to identify individual cells in the sections. In plant tissues the outline of each cell is readily identified by the enclosing cell wall (Fig. 28), but in animal tissues the appressed plasma membranes of adjacent cells often make this very difficult. Within cells the nucleus is readily distinguishable from the cytoplasm, as are large vacuoles in plant cells (Fig. 29). The cytoplasm contains the organelles dispersed in a matrix, whose visible components are mostly free ribosomes, polysomes, glycogen granules, lipid droplets, etc., from which the organelles are delimited by the dark lines of their surrounding membranes.

In animal cells mitochondria are characterised by their two enclosing membranes, with the inner one infolded to form the cristae (Figs. 31 and 32). However, in non-photosynthetic plant cells there can be confusion between mitochondria and plastids of similar size and shape. The latter have darker staining envelope membranes and stroma contents (Fig. 29). In photosynthetic plant cells the presence of granal membrane stacks readily identifies the chloroplasts.

Pairs of parallel membranes, studded on the cytoplasmic face with ribosomes are characteristic of rough endoplasmic reticulum, while such membranes without ribosomes are smooth endoplasmic reticulum (Figs. 31 and 32). In animal cells the smooth membranes of the Golgi apparatus are easily confused with smooth endoplasmic reticulum, but can be identified by their association together in stacks and the presence of peripheral vesicles (Fig. 32). Such problems rarely arise in plant cells, with their discrete units of the Golgi apparatus, the dictyosomes (Fig. 29).

The single membrane-bound vesicles formed by the Golgi apparatus are readily identified in plants, but in animal cells a variety of vesicles, with different origins and functions, may be present. Many are secretory vesicles from the Golgi apparatus, destined for fusion with the plasma membrane, others are part of the lysosomal system in the cell, with incoming vesicles from the plasma membrane (pinosomes and phagosomes).

Microbodies are single membrane-bound organelles that are distinguishable from vesicles by their larger size and irregular margins. Some are also characterised by the staining of the matrix and the presence of dense or crystalline inclusions.

The cytoplasmic matrix also contains a number of tubular and fibrillar structures. Microtubules are characterised by their almost constant size (24 nm diameter) and relatively straight, unbranched cylindrical structure. Organised fibrillar arrays are a feature of muscle cells, but randomly organised fibrils and filaments are also a feature of non-muscle animal cells and plant cells.

From an examination of a collection of micrographs through similar cells it should be possible to arrive at some tentative conclusions about the types of organelle present, their shape and their distribution in the cell.

Three-dimensional interpretation

One of the main problems associated with the interpretation of electron micrographs is the loss of three-dimensional information about the shape and relative dispositions of the cell components, resulting from thin sectioning and viewing a projected image in the electron microscope. The shapes of the resultant two-dimensional profiles may be characteristic of one or more three-dimensional shapes, so some care is needed in their interpretation.

Circular profiles usually originate from spherical structures and oval profiles from ellipsoids, but both could originate from cylindrical structures. In these cases the profiles should be examined to determine whether all their perimeters are uniform, indicating that the structure has been sectioned normally (at right angles to an axis). If the perimeter is not uniform, specifically having one opposite pair of sides discrete and well-defined, while the other pair of sides are diffuse, then the structure has been sectioned obliquely and is cylindrical or an elongated ellipsoid.

A large number of profiles should be examined in as many micrographs as possible so that a complete range of section angles with respect to the components are achieved. This will minimise the risk of misinterpretation. For example, a common mistake is to assume that a micrograph depicting a portion of cytoplasm completely surrounded by an organelle represents an enclosed cytoplasmic enclave isolated from the rest of the cell (Fig. 1). In many cases, sections in other planes reveal that the organelle is cup- or flask-shaped and that the 'internal' cytoplasm is in continuity with the rest of the cell cytoplasm.

Fig. 1. Section of a plastid. In this section the plastid profile appears to contain a cytoplasmic enclave, however it would be incorrect to assume that this portion of cytoplasm was isolated from the rest of the cell. The plastid may be cup- or flask-shaped, an interpretation supported by the obliquely sectioned envelope membranes at one side of the structure (arrowed). *Avena* root tip ($\times 17\,400$).

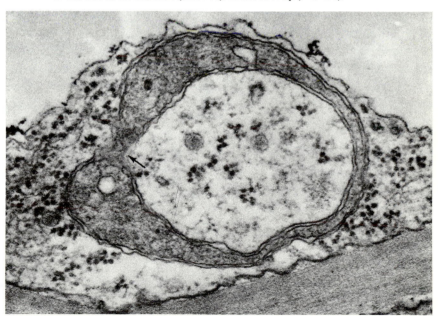

Two methods are available for improving the three-dimensional interpretation of cell structure. One is to cut and record serial thin sections and reconstruct a three-dimensional model of the cell, the other is to cut thicker sections (1–5 μm thick) and examine them in either a conventional (up to 120 kV) or high voltage (up to 1000 kV) transmission electron microscope.

Serial sections

Serial sections require some additional skill in their preparation. A small block face is desirable, so that a large number of sections can be contained on a standard 3 mm diameter grid. Plastic (formvar or collodion) coated slot grids are ideal as there are no grid bars to obscure parts of the sections. All the micrographs should be taken at the same magnification and a set of standard prints prepared.

One cheap, but satisfactory, method of reconstruction uses standard polystyrene insulation boards (Atkinson, John and Gunning, 1974); several other methods have been suggested (e.g. Dunn, 1972; Rakic, Stensas, Sayre and Sidman, 1974; Sandler, 1974; Ware and Lopresti, 1975; Calvayrac and Lefort-Tran, 1976; Jordan and Saunders, 1976; Stevens, 1977). A tracing of each micrograph is prepared on acetate sheets which are projected by an overhead projector on to a sheet of polystyrene, temporarily affixed to a wall. The magnification of the final image is adjusted to correspond to the magnification factor in going from the original section thickness (Gunning and Hardham, 1977) to the thickness of the polystyrene sheet. Each micrograph tracing is projected and outlined on a fresh sheet. The sheets are cut out using a hot wire cutter and assembled to make a model of the cell. Such reconstructions can be very revealing and provide a few surprises, as can be seen by comparing a thin section through a *Chlorella* cell with the polystyrene model constructed from this and other sections of the same cell (Fig. 2).

Thick sections

Thick sections attempt to provide a solution to the problem of three-dimensional interpretation, by retaining enough of the original structure in the section so that its shape, and relationships to other components, can be observed. Frequently pairs of photographs are taken from the same area of the section, but at different tilt angles relative to the electron optical axis, so that the resulting micrographs form a stereo-pair giving an impression of depth in the recombined image.

There are problems in obtaining the transmission of sufficient electrons through such thick sections. At conventional voltages (up to 80 kV), transmission of electrons is low and the specimen temperature rises due to the absorption of a high proportion of the energy from the electron beam. This led to the development of instruments capable of accelerating electrons to 1000 kV and more (Fig. 3). While these instruments are useful to biologists, their general inaccessibility for routine work, and the problems involved in observing low-

Fig. 2. Three-dimensional reconstruction of *Chlorella* cell. (a) Thin section of a *Chlorella* cell, with nucleus (N), chloroplast (C) containing a pyrenoid (P), microbody (M) and mitochondrial profiles (arrows). This is one of a set of serial sections through the same cell ($\times 15\,750$). (b) Model constructed from polystyrene sheets traced from the set of serial sections. The nucleus has been removed to reveal the centriole pair. The single branched mitochondrion ramifies throughout the cell, and through cytoplasmic channels penetrating the chloroplast. (a and b, courtesy of Atkinson, John and Gunning.)

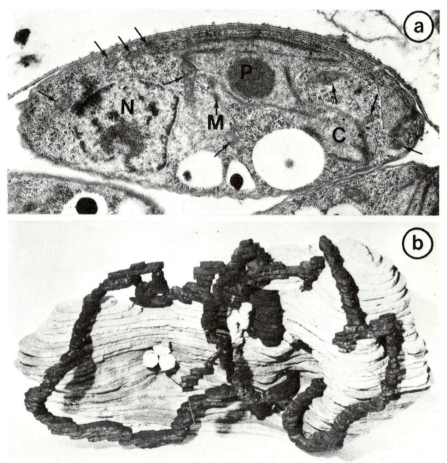

contrast specimens through thick lead glass viewing ports, have not encouraged this approach. One problem encountered in conventional transmission microscopy is the severe chromatic aberration resulting from attempting to form an image with electrons possessing electron energies (equivalent to wavelength) varying from 80 kV down to very low levels. By raising the accelerating voltage slightly (to 120 kV) and screening out electrons with lower energies more efficiently it has been possible to produce conventional

Fig. 3. Thick section of a tapetum cell viewed in a high-voltage electron microscope at 1000 kV (AE1 EM7). The cytoplasm contains branched, elongate vesicles which only appear as circular to elliptical profiles in their sections. The nucleus lies to the left of the figure and there is extracellular space (*) between this cell and the adjacent cell to the right of the figure. Anther of *Avena sativa* (×26 700).

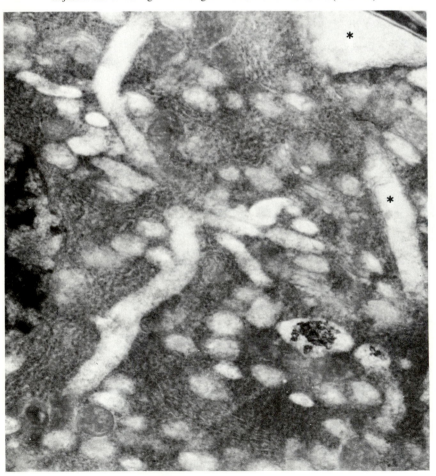

transmission electron microscopes useful for sections $1-2\ \mu m$ thick (Fig. 4).

Conventional staining techniques introduce too many heavy atomic nuclei into thick sections, resulting in a very confused image formed by overlapping cell components. Stains are being developed that selectively bind to the organelle under investigation (Fig. 5), so that it is revealed against a relatively unstained background (Poux, Favard and Carasso 1974; Thiéry and Bergeron, 1977; Harris, 1978; Thiéry, 1979).

Functional interpretation

The qualitative analysis of a set of electron micrographs usually leads to a discussion of the possible functions and roles of the components and cells

Fig. 4. Thick section of tapetum cell, similar to that described in Fig. 3, viewed in a conventional transmission electron microscope at 120 kV (Philips 400). The three-dimensional organisation of the elongate vesicles is again visible. The darkly stained organelles are plastids with electron lucent channels in their interiors ($\times 21\ 250$).

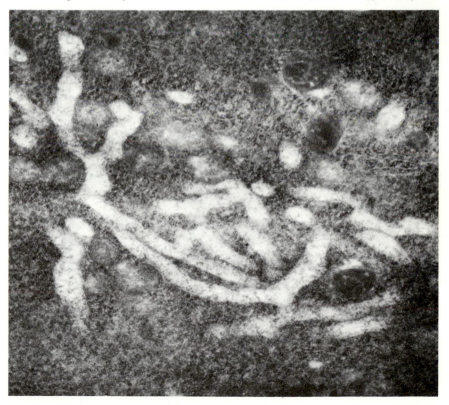

Such discussions are based on the previously established activities of similar components in other cell types. The abundance (either in terms of apparent numbers or size) of particular cell components and their associations with each other, or with particular regions of the cell, are taken as significant factors in postulating their activities and that of the cell.

There are obvious dangers in this approach, but used with caution it can become an invaluable guide to aid the planning of future work, designed to establish the roles less ambiguously. It should be possible to avoid the most obvious pitfalls. For example, there is little value in postulating a role for an organelle that is quite outside its established chemical activities and capabilities, unless there is strong evidence for doing so.

Some dynamic activities of living cells involve movement and changes in structure which are halted by the fixation process. This results in a range of component morphologies and positions in the cell, preserved in the electron micrographs. Sometimes a complete sequence can be found in a single micrograph, as in the case of Golgi vesicles, where stages of their formation on cisternae, transport through the cytoplasm and fusion with the plasma membrane can all be found on a single micrograph. Other sequences, such as chromosome movements during mitosis, can only be built up from a collection of micrographs.

While such images can be objectively assembled in a logical sequence it is, of course, important to remember that there is often no information available as

Fig. 5. Increased contrast of cell components. The section of *Avena* leaf epidermal cell was stained by the Thiéry method, this results in a heavy deposition of osmium in many of the dictyosome cisternae and vesicles ($\times 37\,500$).

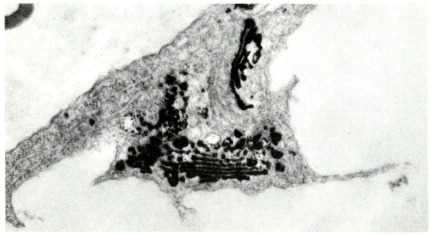

to the direction in which the sequence was running in the cell at the time of fixation. For example, vesicles believed to be fusing with the plasma membrane and releasing their contents outside the cell could equally be pinosomes entering the cell.

PART III

Quantitative analysis

Introduction

The general conclusion from numerous studies on all types of cell, one that is expounded in every cell biology text, is that the structure of a cell is intimately related to its function. It follows from this that the *capacity* of a cell to carry out a particular function is limited by the *amount* of the relevant components and structures present. This is recognised in a qualitative fashion by such phrases as: the cell has numerous mitochondria; a well-developed endoplasmic reticulum is present; there is low Golgi activity at maturity. Clearly these descriptive terms lack precision, they are of little use when comparing different cells, or similar cells growing under different conditions. They are even less useful when attempting to assess the potential of a cell with respect to a particular function, such as the rate of product formation.

Cellular reactions depend on enzymes. These are found either bound to cell membranes or 'free' in the cytoplasm. Bound enzymes possessing related functions are frequently found associated together on specific and readily identifiable membranes, e.g. mitochondrial cristae. Similarly many 'free' enzymes are associated in compartments forming the recognisable groups of organelles: mitochondria, plastids, microbodies, etc. The *density* of bound enzymes on their supporting membranes and the *concentration* of 'free' enzymes in their internal compartments are parameters which determine the total productive capacity of an organelle. These values may be difficult to determine, however they are often proportional to the total membrane surface area and compartment volume respectively.

Determination of this productive capacity by biochemical means is far from straightforward. If the tissue is not homogeneous the required cell type must first be separated from the surrounding cells. Organelle content can be estimated by cell fractionation to yield some measure of the amount of each present. Fractionation techniques are rarely very efficient, with a good deal of cross contamination and simple loss of components occurring so that 100% recoveries are not possible. Some organelles, especially the endoplasmic reticulum, are so complex and morphologically diverse that it is just not possible to use fractionation methods to separate the functionally different parts effectively.

Electron micrographs, on the other hand, provide excellent discrimination between different cell types in a tissue and between organelles within each cell. This allows unequivocal identification of both the desired cell type and the organelles present. The membranes and compartments can be identified and quantified to far finer levels than by using cell fractionation methods, for example, one can distinguish smooth endoplasmic reticulum elements from contiguous rough elements, inner from outer mitochondrial membranes, stroma lamellae (frets) from thylakoids in chloroplasts, etc. The following sections detail the methods that are available to the microscopist for the quantification of cell structure.

Direct methods of quantitative analysis

Linear dimensions

Much basic information on cell structures can be acquired directly by measuring dimensions on micrographs. The accuracy of these measurements will depend on the stability of the sections, the calibration of the electron microscope and the accuracy of the photographic procedures leading to the final print (see p. 67). The final measurements from the micrograph are usually made in millimetres and these should be converted to micrometers (multiply by 1000, or 10^3) or nanometers ($\times 10^6$) before division by the print magnification to avoid simple arithmetic errors in calculating the true dimensions. Many investigators prefer to measure the sizes of the objects directly on the negative, using a travelling microscope.

Table 1. Surface areas and volumes

	Surface area	Volume
Cube	$6L^2$	L^3
Sphere	$4\pi R^2$	$\frac{4}{3}\pi R^3$
Cylinder	$2\pi R (R + h)$	$\pi R^2 h$
Ellipsoid	$4\pi ab$	$\frac{4}{3}\pi ab^2$

Surface areas and volumes

Linear dimensions from thin sections can be used to determine surface areas and volumes of cells and components where these have a regular shape (Table 1). Direct measurement of the major axes can be made with the light microscope using whole cells or sections made in the appropriate planes. It is always advisable to convert these initial measurements into true dimensions first since errors are easily made in reducing area or volume measurements by the magnification factor to give true cell values.

Cellular components with similar regular shapes can be treated in the same way, though there is more room for error in deciding whether the structure has been sectioned in a particular plane. Perhaps the commonest regular shape encountered is a sphere, as with many nuclei and vesicles. Thin sections through a population of identical spheres will yield a series of circular profiles, the largest corresponding to median sections through the sphere diameter. Measuring the largest profile is certainly a simple way of determining the radius and hence the surface area and volume of the sphere. However, in biological situations, it would be unusual to find a population of structures with identical dimensions (viruses are an obvious exception), usually there is a distribution of sizes about some mean value. Hence the largest circular profile obtained on sectioning may only be representative of a small fraction of the spheres in the population. So direct measurement is not recommended; later we will discuss other, indirect, methods for estimating the sizes of spheres (p. 32).

Numbers

If structures are too small to be counted in the light microscope then some attempt will have to be made to obtain the information from thin sections. Here the problem is different, the structures are usually readily visible, but are often so much larger than the thickness of the section that it is impossible to be certain of their number in the whole cell from a single micrograph. The dimensions of the structures determine the method to be used for estimating their numbers, apart from the method of cell reconstruction described previously (p. 14). The numbers of structures larger than the section thickness and of unknown disposition in the cell are difficult to determine and will be discussed later. The remaining structures may be contained wholly within the section or pass through it.

Structures contained within the section thickness can be directly counted in the micrographs. Numbers of particles per unit area of micrograph can be expressed in units of cell volume if the section thickness is known (p. 67). If the particles are very numerous, ribosomes for example, then 'quadrats' of transparent acetate sheet, cut to correspond to, say, one micrometre square allowing for the micrograph magnification, are placed randomly over the micrograph and the mean number estimated.

Where the structures are restricted to a particular surface in the cell, such as nuclear pores, plasmodesmata in plant cell walls or junctional complexes between animal cells, then numbers per unit area can be counted on tangential, or grazing sections. Extrapolation to the whole surface to give total numbers can be carried out if their distribution is uniform over the whole surface, which is sometimes not the case for nuclear pores and often not true for plasmodesmata. Since nuclear pores are regular structures with identical diameters in a given cell type (at least), it is possible to estimate their density on the surface from sections through the nuclear envelope. The largest pore profile will correspond to the diameter of the pore so giving its surface area; while the proportion of the surface occupied by pores can be estimated by measuring the proportion of the sectioned envelope profile (P) occupied by pores (using a thread and ruler respectively). Then number per μm^2 equals P / Pore area in μm^2.

Structures passing through the section can be quantified if they have a regular disposition, for example, the parallel fibres of muscle and many microtubular systems. Here the total length of the system can be determined and the number of structures counted in sections at right angles to their long axes.

Indirect methods of quantitative analysis: stereology

The size, shape and distribution of most cell components preclude the estimation of their size, numbers and total contribution to the cell by direct observation of whole cells or thin sections, but it is clear that the frequency of their occurrence in thin sections must be related to the level of their presence in the whole cell. Stereology is a means of quantifying this relationship, providing estimates of the volumes, surface areas and numbers of components within the cell.

The following sections set out the basic stereological principles and methods for making these estimates. Used wisely they can provide an amazingly detailed insight into the workings of cells, but abuses will lead to erroneous conclusions. Careful sampling of the tissue and suitable statistical safeguards must be employed, so the reader should not ignore the sections on these subjects.

The methods involve certain numerical relationships between various parameters. The stereological parameters have been assigned specific symbols (Table 2), which will be used sparingly in this text.

Volumes

If a component occupies a large part of a cell in a micrograph, either by virtue of its numbers or size, we are inclined to believe that it also occupies a large part of the cell. This interpretation is a purely subjective one and is affected by the size, shape and distribution of the individual components across the face of a micrograph. It is also a particularly insensitive method of assessment, requiring relatively large changes in the components before any change is noted by the observer. The basic premise of this subjective assessment is correct and was first demonstrated by Delesse in 1847 working on polished flat faces of conglomerate rocks. He concluded that the mean surface area proportions (A_A) on two-dimensional faces are the same as the mean volume proportions (V_V) of the components within the rock (Weibel, 1969; Weibel and Bolender, 1973).

$$A_A = V_V$$

The mathematical proof for this relationship will not be given here, instead a simple illustration will be provided deriving the same relationship for thin sections. A thin section has a finite thickness and hence the volume of all the components in the section can be determined by multiplying their surface areas on the section face by the thickness (Fig. 6). Since all the thickness terms are identical it follows that the ratios of their volumes are the same as the ratios of their surface areas. An individual section represents a random sample of the tissue, by preparing many tissue blocks and examining a single section from each it is possible to obtain a mean surface area ratio for each component with respect to all the other components and the total section area that contains them. A very important reservation stems directly from this model, it assumes that the surface area of an individual component is the same on both faces of the section, with the component boundaries at right angles (normal)

Table 2. Mathematical symbols in stereology

Symbol	Definition	Dimensions
A_A	Area fraction occupied by component	$\mu m^2/\mu m^2$
V_V	Volume fraction occupied by component	$\mu m^3/\mu m^3$
L	Length of test line	μm
L_T	Total length of test line on grid	μm
I	Intersection of surface boundary with test line	μm
S_V	Surface density of component	$\mu m^2/\mu m^3$
P_A	Density of linear components passing through test grid area	μm^{-2}
L_V	Density of length in unit volume	$\mu m/\mu m^3$
N_A	Number of profiles per unit test area	μm^{-2}
N_V	Number of particles per unit volume	μm^{-3}

to the section plane. These faces cannot be separated on the resultant micrograph. For most structures this does not matter, they are much larger (1–10 μm diameter) than the section thickness (50 nm). Later we shall discuss correction factors that should be applied when small components are encountered (p. 67).

The problem of determining relative component volumes is now reduced to one of determining surface areas on micrographs. This could be done, laboriously, by superimposing a piece of graph paper on the micrograph and counting all the small squares lying over each component, giving accurate values for a single micrograph. But we need to examine a sufficiently large random collection of micrographs to avoid sampling errors, and in practice a considerable variation is found in the values from one micrograph to the next. Since we are only interested in relative surface areas of components we can obtain an estimate by reducing each graph paper square to a point, so generating a square lattice of points. Counting the relative number of points over each component and over the whole micrograph will give an estimate of the surface areas of the components relative to each other and the whole surface (Fig. 7). The loss in accuracy in recording the values from individual micrographs is insufficient to reduce the accuracy of the final estimate. So volume determinations in cells are relatively easily made by counting points over micrographs of thin sections.

Fig. 6. Relationship of volume fractions to area fraction. (a) A diagram of an edge-on-view of a section of thickness *t*. The shaded portions represent the edge-on-views of the components seen in (b), a top view of the section. All the components have the same thickness *t*, so their volume depends on the area they occupy in (b).

a

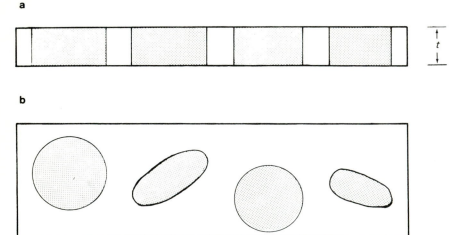

b

Surface areas

Membrane surfaces in cells are analogous to boundary interfaces between phases in a solid, and hence their areas can be determined by methods devised for such mixed phase solids. The methods determine *surface density* values (S_V), that is, surface area per unit volume ($\mu m^2/\mu m^3$), where the value refers to the interface surface area. These values are used directly when referring to the three-dimensional structure of a membrane, ignoring the important fact that a membrane has two distinct, and probably functionally different, surfaces. The true surface area is twice the determined interface surface area.

Although the method for estimating surface area is simple, involving a straightforward counting procedure as in volume determinations, the basis for this method is less obvious. Again we can subjectively examine a micrograph and ascertain whether the membrane levels in a whole cell are high or low, so that the amount of membrane present in a micrograph is related to the level in the whole cell. If we rule, or superimpose, a straight line (a test line) across the micrograph we see that it intersects with (crosses) the membranes. The more membranes that are present the greater the number of intersections, and we can quantify this by counting. It has been shown that the surface

Fig. 7. (a) Assessment of the area fraction occupied by two components, A and B with a grid of 5-mm squares. The total area of the grid is 25 cm², the area of A is 4 cm², and of B is 5·4 cm²; giving area fractions of 0·16 (16%) for A and 0·22 (22%) for B. Each square can be replaced by a central point, as in the bottom left-hand corner of the grid.

(b) As (a) but all the squares replaced by points. The total number of points is 100, 16 lying within the boundary of components A and 25 within components B; giving area fractions of 0·16 (16%) and 0·25 (25%) respectively.

(a)   **(b)**

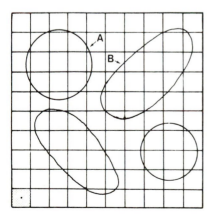
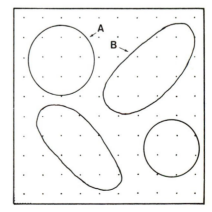

density (S_V) is equal to twice the number of intersections (I) divided by the total line length (L_T).

$$S_V = 2I/L_T$$

If the line length is reduced by the magnification factor of the micrograph (i.e., reduced to real cell dimensions) and expressed in microns, the surface density obtained will have the units $\mu m^2/\mu m^3$.

This relationship has been derived on several occasions, but these derivations require some formal mathematical background to understand them. Weibel (1974) has given a derivation that can be understood without this mathematical background and it will be described here.

A sphere, radius R, is contained within a cube, side L (Fig. 8). An array of uniformly spaced test lines, total number n, length L, runs from one face straight across to the opposite face of the cube, with a number of them (n_0) passing through the sphere. The proportion of these lines that intersect the sphere (n_0/n) is equal to the ratio of the area of the largest cross-sectional area in the sphere (πR^2) to the area of one face of the cube ($\pi R^2/L^2$). The number of intersections with the sphere surface per unit length of test line is twice the number of intersecting lines ($2n_0$) divided by the total line length or by the number of lines times their unit length ($2n_0/nL$). The ratio n_0/n is already

Fig. 8. Derivation of surface density relationship. (a) A sphere, radius R is contained within a cube, edge length L. Test lines, length L, pass perpendicularly through the cube from top to bottom, some penetrating through the sphere. The outline of the sphere is projected on to the bottom face of the cube.

(b) The bottom face of the cube, with the test lines seen end-on and the circular outline (radius R) of the projected sphere.

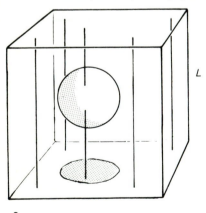

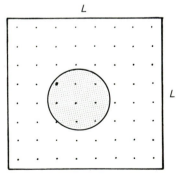

a   b

known to be $\pi R^2 / L^2$, so the last ratio obtained can be rewritten as $2\pi R^2 / L^3$. This value is half the surface density relationship for the sphere in the cube $(4\pi R^2 / L^3)$, and hence the total surface density is given by twice the total number of intersections divided by the total line length.

Lengths

Many biological systems contain filaments, fibres or tubules. Each particular structure usually has a constant diameter, but is often of variable length, so the total content of these structures in the cell or tissue depends on the total length present. Examples of these structures include actin and myosin arrays in muscle, microfilament (actin) and other filamentous arrays in cytoplasm, microtubules in dividing and non-dividing cells and extracellular filaments, such as collagen fibres in animal tissues and microfibrils in plant cell walls.

The mean length per unit volume of these structures can be very easily determined from their numerical density per unit area of the micrograph, corrected for the magnification (Weibel, 1969; Weibel and Bolender, 1973), however, the three-dimensional organisation of the components within the volume must also be determined. This is because the relationship between length per unit volume (L_V) and counts per unit area (P_A) for a randomly oriented system is

$$L_V = 2P_A$$

while for an organised system of parallel fibres sectioned at right angles to their long axes, it is

$$L_V = P_A$$

The second equation would be used for microtubule arrays in a mitotic spindle, or actin in muscle fibres, but the first for collagen fibres in tissues or actin filaments in the cytoplasm of amoeboid cells. Partially oriented systems can be tackled by using the correct multiplication factor for the counts per unit area, but it is simpler to make observations on a series of sections, each from blocks sectioned at several, random, planes through the tissue. This approach would of course provide a suitable safeguard whenever the first equation was being used.

Numbers and sizes

Where comparisons are being made between cells at different stages of development, or growing under different conditions, it is often necessary to compare the size and numbers of the individual components present. This is because total volume and surface area values can obscure changes in the number, size and shape of individual components. A further important reason for determining their number and size is that it allows the calculation of absolute values per cell or tissue from the relative values obtained from stereology. Direct methods of making these determinations have been considered (p. 25), now we will deal with the more usual situation, requiring an indirect approach. Despite the problems that will become apparent during the discussion, it is worth emphasising that these methods are frequently the only ones that enable such information to be obtained.

The methods fall into categories depending on the shape of the component.

Fig. 9. The effects of section thickness on the image. Above: an edge-on-view of a section, thickness t, through a number of membrane bound spherical components with diameters ranging from that of the section thickness (a and b) to four times the section thickness (e and f).

Below: projected view of the section showing the circular profiles. When the component contents are lightly stained (low contrast) their appearance depends on the membrane staining. Small cap sections (b and d) may not be detectable. The interior of the component may be completely obscured (a and c) or partly visible (e and f). If volume fraction determinations are based on counting points within the membrane boundary this will lead to underestimates in e and f and no counts for a and c. Alternatively counting to the outer edge of the component will lead to overestimates in all cases except b and d. If the component contents stain heavily all profiles will be visible and appear uniformly dark to their outer edge (not like e and f), and their volume fractions will all be overestimated.

The membrane presence will be greater in the projected image than in a true infinitely thin section. Compare the number of intersections with the top section face (8), with those at the bottom face (6) and a line through the centres of the circles in the projected image (12). If the cap sections b and d are missed this last value would be 8.

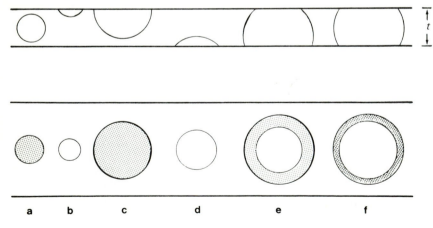

a b c d e f

If the shape is regular, predictions can be made about the relationship between the internal volume and the boundary, or surface, of the component. We have already seen that volumes and surfaces can be estimated. The less regular the shape the more difficult it becomes to make accurate predictions. We will start by discussing spherical structures and move on to less regular shapes later.

Spheres

Certain cellular components approximate to spherical shapes, notably the nuclei and nucleoli of many cells, vesicles and lipid droplets. As the nucleus is usually present as a single entity in the cell, the determination of its size can lead directly to an estimate of the total cell volume, from the volume fraction of the cell occupied by the nucleus.

There are two basic methods for estimating sphere size and number. The first depends on the relative frequency with which median and non-median sections are cut through a sphere of a given size, while the second depends on the relationship between the enclosed volume and its bounding surface.

Sphere size from section profiles. A population of identical spheres, sectioned at random, will yield a population of circular profiles from the largest, corresponding to equatorial sections, to the smallest, corresponding to a tangential cap contained in the section thickness (Fig. 9). The size distribution of circular profiles obtained by cutting random sections through a collection of such spheres is quite predictable (Fig. 10). In practice, biological structures are rarely 'identical' but are often 'similar', that is, there is a more or less normal distribution about a mean size. Further, small tangential sections of spherical structures are of low contrast, and so may be missed in the micrographs. Thus faced with a micrograph containing sectioned vesicles, nuclei or other spherical structures there is the problem of determining whether there are one or more size classes of spherical structure present and the extent to which the smallest profiles are missed from the size distribution.

If the spheres are of identical size, then the largest profile diameter would correspond to the sphere diameter. This is, as already stated, unlikely, so the largest profile would be quite unrepresentative of the whole population. A simple estimate of the mean sphere diameter can be obtained from the following:

$$\bar{D} = 4\bar{d}/\pi$$

where $\bar{D}$ is the mean diameter of the spheres and $\bar{d}$ is the mean diameter of the circular profiles in the sections (Weibel, 1969; Weibel and Bolender, 1973). It suffers from the disadvantage that the missed small profiles lead to an overestimate of $\bar{d}$. Plotting a distribution diagram of the profile sizes may enable the smallest size class to be estimated by interpolation and so improve the accuracy of $\bar{d}$, but this is not satisfactory (Greeley & Crapo, 1978).

Giger–Riedwyl method for sphere size. Plotting a histogram of the profile size distribution forms the basis of a method developed by Giger and Riedwyl (1970; set out in detail in Weibel and Bolender, 1973) for the mean sphere diameter, $\bar{D}$, and the standard deviation of the population. In effect this method determines a correction factor for $\bar{d}$ that is more appropriate for the

Fig. 10. Profile distribution from sectioning a sphere. The circle represents a sphere with a diameter of 1·0. Sections through the sphere yield circular profiles ranging in diameter from 1·0 to nearly zero. Here the diameters of four such circular profiles are represented by vertical lines, with values of 1·0, 0·9, 0·8 and 0·5 respectively. These circular profiles are formed by sectioning the sphere at different distances from the sphere centre, measured along the radius of the sphere.

   The histogram is a plot of probability against the possible circular profile diameters between 0 and 1·0. This continuous range of diameters is divided into ten size classes, each of 0·1 unit, for convenience. The height of each histogram block represents the probability of obtaining a circular profile with a diameter within the range of sizes covered by that block when cutting random sections through a sphere of size 1·0.

   The height of each histogram block is proportional to the distance, along the sphere radius, between the circular profile diameters defined by the boundaries of the block. These become successively smaller further from the sphere centre, for example, the distance, x, between diameters 1·0 and 0·9, is larger than that between 0·9 and 0·8 and so on.

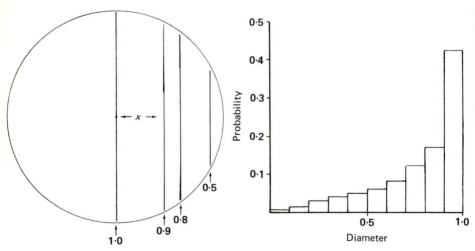

sphere population under consideration than $4/\pi$ ($= 1.27$). The method is suitable for normal distributions of sphere sizes, but does not take into account the effect of section thickness (p. 67). As previously the diameter of a large number (200–500) of profiles are measured on random sections (Fig. 11). A frequency histogram is constructed composed of 10 to 16 size classes, of equal size intervals from the smallest to the largest, starting at a value of half the size class interval (Fig. 12). Clearly the graph of the distribution should go through the origin, so the smallest size classes, which often contain no counts or are underrepresented due to missing small profiles, are estimated by interpolating points at the means of each class between the origin and the nearest complete size classes. Errors introduced by completing the histogram in this way can be estimated and corrected later. The mean profile diameter, $\bar{d}$, is calculated from the completed histogram and the first estimate obtained of the sphere diameter, $\bar{D}_1$, as before:

$$\bar{D}_1 = 4\bar{d}/\pi$$

Fig. 11. Nuclear size in *Chlorella* cells. Thin section of a *Chlorella* cell selected to show the nucleus (N), slightly flattened on one side adjacent to the edge of the dictyosome (D). The flattening is more pronounced in sections which pass through the nucleus and the middle of the Golgi apparatus. ×25 500 (courtesy of A.W. Atkinson).

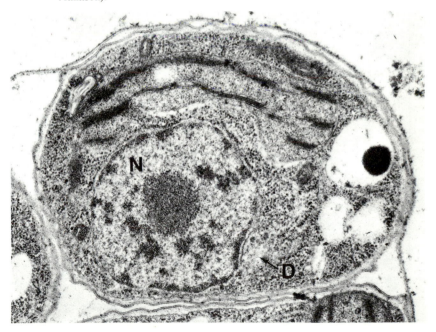

The largest profile diameter corresponds to the largest sphere diameter, so the end of the histogram defines the limit of the distribution. For a normal distribution, the mean ± three times the standard deviation ($\pm 3\sigma$) includes 99·73% of the population. So for the profile distribution the approximate standard deviation is one-third of the difference between the value $\bar{D}$ and the

Fig. 12. Size distribution of *Chlorella* nuclear profiles. The square root of the product of the major and minor axes of each nuclear profile was taken as the profile diameter (d). A total (N) of 149 profiles were measured and plotted to form the histogram. The class size (a) is 0·2 μm, with class number i $= 1$ running from 0·1 to 0·3 μm, i $= 2$ from 0·3 to 0·5 μm and so on. The small size classes are assumed to have been underestimated and have been increased (dotted lines) from 1 to 3 for the 0·4 μm class and 6 to 9 for the 0·6 μm class. The mean profile diameter is 1·38 μm, and the first estimate of the sphere diameter is 1·76 μm. The ratio (P) of the area of the histogram above 1·76 μm to the area of the whole histogram is 0·150. From Fig. 13 the corrected value for $F(P)$ is 1·267, giving a second estimate of the nuclear size of 1·75 μm. The calculated number of profiles that should be in each of the small size classes are 3·6 in class 0·2 μm, 7·2 in class 0·4 μm and 10·8 in class 0·6 μm.

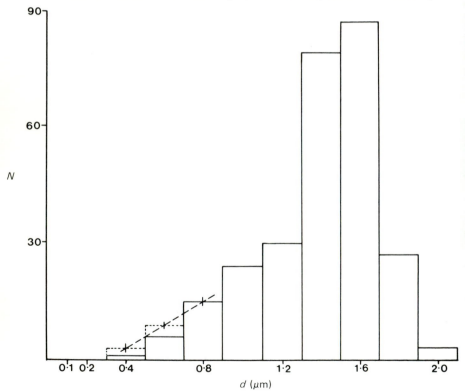

largest profile diameter. To obtain the correct estimates of sphere diameter and standard deviation, the area of the histogram above the value of $\bar{D}$ is divided by the area of the whole histogram to obtain the value P. From the graphs in Fig. 13 the two correction factors for this value are determined, $F(P)$ and $f(P)$. This enables a second estimate of the mean sphere diameter, D_2, to be calculated:

$$\bar{D}_2 = F(P)\bar{d}$$

and the standard deviation by

$$\sigma = f(P)\bar{d}$$

The accuracy of the earlier interpolation can now be determined. Where N is the total number of profiles in the completed histogram, the number of profiles (N_i) in each of the smallest classes is

Fig. 13. Coefficients for profile mean diameters, $F(P)$, curve a, and standard deviation, $f(P)$, curve b, plotted against P. Each value of P yields two values, one from each curve, read from the appropriate axes (redrawn from Weibel and Bolender, 1973 after Giger and Riedwyl, 1970).

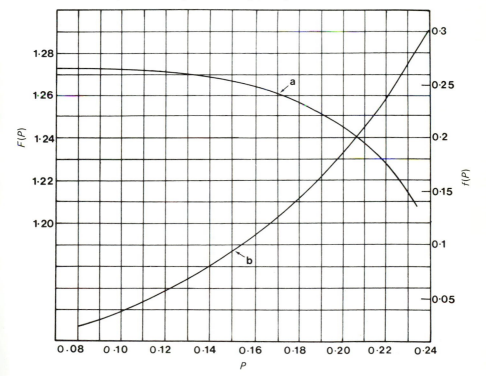

$$N_i = ia^2N/\bar{D}_2^{\ 2}$$

where i is the class number ($\frac{1}{2}$, 1, 2, 3 etc.) and a is the size of the class interval (with the same units as $\bar{D}_2$). If the numbers in each calculated size class differ markedly from those found by interpolation, then the whole exercise is repeated, using these calculated numbers to complete the original histogram.

The mean sphere diameter, $\bar{D}$, can be used to find the number of spheres per unit volume (N_V) of the cell or tissue from

$$N_V = N_A/\bar{D}$$

where N_A is the mean number of circular profiles per unit area of section. The value obtained should correspond to that which can be calculated by dividing the mean sphere volume into the volume fraction of spheres present per unit volume of cell or tissue determined by point counting (p. 28).

Sphere populations of different sizes. Distribution diagrams of profile sizes are invaluable for testing the heterogeneity of the sphere population. Indeed, by comparison with the standard curves published by Hennig and Elias (1970) it is possible to relate the profile size distribution to the distribution of sphere sizes present. These curves emphasise the care which must be taken in interpreting information gained from measuring the diameters of circular profiles in electron micrographs, however they have not been widely used in biology.

Several methods exist which yield numerical densities of each sphere size class present. Again these are based on the expected distribution of circular profile diameters from random sections through a sphere. The first method specifically developed for microscope sections is that by Wicksell (1925); it has been successfully used by Baudhuin and Berthet (1967) for determining size distributions of approximately spherical isolated mitochondria in a centrifuged pellicle. Their paper includes a detailed discussion of the methods and the practical problems encountered. Methods developed for polished faces of materials are applicable to biological situations, but they should be used with caution since they do not allow for section thickness effects. One of the most useful methods is an adaptation by Saltykov of a method devised by Johnson (see Underwood, in DeHoff and Rhines, 1968), which is similar to the Wicksell transformation. Essentially it involves dividing the profile diameters into a series of size classes on a logarithmic scale, and assuming that those in the largest size class represent median sections through the largest spheres. Tables are then used to give an estimate of the contribution of all non-median sections from this particular size class of sphere to the remaining, smaller, size classes of circular profiles. These are deducted from the counts for each of these classes and the whole process repeated, taking the next profile diameter size class down as representing median sections through the next lower size class of sphere, and so on.

Saltykov method. A modification of this method by Saltykov, described by Underwood (1970), is less cumbersome and does not require the use of extensive tables and calculations. It depends on dividing the areas of the profiles into class intervals on the basis of the ratios of profile area (A) to area of the largest profile (A/A_{max}). As with the previous method, it is dependent on the largest profile size (A_{max}) to determine the numerical density of the largest sphere population. The class intervals for the ratio A/A_{max} are based on a log scale of diameters with the factor $10^{-0.1}$ ($= 0.7943$), with a maximum number of 12 classes (Table 3).

The diameters (d) of a large number of profiles (> 300) are measured on random sections and divided into size classes according to the diameter ratios in Table 3, such that the first class contains all those from the largest (d_{max}) to $0.7943 \times d_{max}$, the second class from $0.7943 \times d_{max}$ to $0.6310 \times d_{max}$, etc.

The general working formula for finding the numerical density (N_V) is as follows:

Table 3. *Saltykov area analysis: group limits and corresponding frequency of sections for a spherical particle*

Group number	Relative section diameters, d/d_{max}	Relative section areas, A/A_{max}	Per cent of sections per unit area, N_A
1	1·0000	1·0000–0·6310	60·749
2	0·7943	0·6310–0·3981	16·833
3	0·6310	0·3981–0·2512	8·952
4	0·5012	0·2512–0·1585	5·200
5	0·3981	0·1585–0·1000	3·134
6	0·3162	0·1000–0·0631	1·926
7	0·2512	0·0631–0·0398	1·195
8	0·1995	0·0398–0·0251	0·747
9	0·1585	0·0251–0·0158	0·469
10	0·1259	0·0158–0·0100	0·294
11	0·1000	0·0100–0·0063	0·185
12	0·0794	0·0063–0·0040	0·117
·	·	·	·
·	·	·	·
·	·	·	·

$$(N_V)_i = \frac{1}{D_i}[1.6461(N_A)_i - 0.4561(N_A)_{i-1} - 0.1162(N_A)_{i-2}$$
$$- 0.0415(N_A)_{i-3} - 0.0173(N_A)_{i-4} - 0.0079(N_A)_{i-5}$$
$$- 0.0038(N_A)_{i-6} - 0.0018(N_A)_{i-7} - 0.0010(N_A)_{i-8}$$
$$- 0.0003(N_A)_{i-9} - 0.0002(N_A)_{i-10} - 0.0002(N_A)_{i-11}]$$

where i is the size class number, starting with the largest $i = 1$; $(N_V)_i$ is the numerical density of spheres D_i in size class i; $(N_A)_i$ is the number per unit area of section profiles of size class i. Any derived size class can be calculated in any order, the terms of the formula are only used down to $(N_A)_1$, so the whole formula is only used for the smallest size class $(i = 12)$.

An example is given in Table 4 for vesicle size distributions in a pollen tube (Fig. 14), each numerical density was derived from the working formula presented above. For example the fourth size class figure was calculated as follows:

$$(N_V)_4 = \frac{1}{0.167}[1.6461(1.6087) - 0.4561(1.2374)$$
$$- 0.1162(0.4563) - 0.0415(0.0851)]$$
$$= 12.77$$

Table 4. *Vesicle sizes and frequencies*

Group Number	Diameter of sections (nm)	Number of sections per μm^2	Mean class Diameter of vesicles (nm)	Number of vesicles per μm^3
1	333–265	0.085	299.0	0.42
2	265–210	0.456	237.5	2.69
3	210–167	1.237	188.5	8.76
4	167–133	1.609	150.0	12.77
5	133–105	1.601	119.0	15.25
6	105–84	1.199	94.5	13.20
7	84–66	0.240	75.0	0
8	66–53	0.085	59.5	1.87
9	53–42	0.008	47.5	0
			Total	54.96

Vesicle sizes and densities in pollen tube cytoplasm (*minus* vacuoles) after 10 minutes treatment with 0.3 $\mu g/ml$ cytochalasin D. A total of 843 vesicle profiles were measured (courtesy of J. Picton).

Note that the units for (N_V), (N_A) and D should correspond (in this case μm). If an automated system is available for determining profile areas directly, then these can be used in the same way (Table 3), although setting memory stores to the correct size class boundaries is not possible until A_{max} is known. Such systems are especially useful when the profile boundary is distorted.

Sphere distributions corrected for section thickness. The methods discussed above were developed for use on plane polished surfaces. These are not ideal for transmission microscopy images since the profile diameters measured on

Fig. 14. *Tradescantia* pollen tube after 10 minutes treatment with 0·3 μg/ml cytochalasin D. Numerous vesicles are present in the cytoplasm of the growing tube. They are probably formed by the dictyosomes (D) and normally move to the tip where they fuse with the plasma membrane, elongating the tube. Vesicle movement and tip growth is inhibited by cytochalasin D, leading to changes in the number, size and distribution of the vesicles ×26 000 (courtesy of J. Picton).

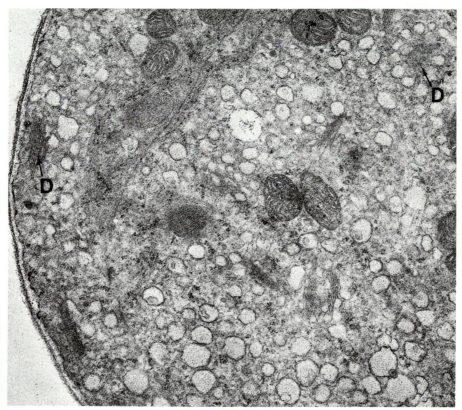

micrographs from sections are always larger than those from a flat plane (Fig. 9). Rose (1980) has established general working formulae for determining sphere size distributions corrected for section thickness. His tables for ten size classes and a range of section thickness are reproduced here (Tables 5 and 6).

As before, a large number of profile diameters are measured from a known area (A) of sections of known thickness, T (p. 68). The section thickness is used to establish ten class sizes that covers the profile distribution (the largest profiles in the highest class numbers) such that the ratio of the class interval (Δ) to section thickness (T) falls into one of those given in the left hand column of Table 5. The number of profiles (n) in each class (1, 2, 3 ... k) is determined and used to find the numbers of spheres (N) with a mean diameter at the midpoint of each sphere class (1, 2, 3 ... m) from,

$$N_m = \frac{1}{A\Delta}[P_{mk}][n_k]$$

where A is the area, Δ the class internal size, both in the same units. N_m is then the number per unit volume. Table 5 is a conversion matrix for $[P_{mk}]$ and should be used as follows:

(i) select the part of the table appropriate for the chosen ratio T/Δ.

(ii) start with the smallest size class and multiply the number of profiles in that size class by the value listed in row 1, column 1, in that part of the table. Subtract from this value the products of each successive table number, across the first row, with the corresponding number of profiles in that size class. The results for the smallest sphere size classes may be negative values, in which case they should be regarded as zeros.

(iii) repeat the whole calculation for each sphere size class in turn, using successive rows of the table.

(iv) the values obtained are multiplied by $1/A\Delta$ to give the number per unit volume for each sphere size class.

The total number of spheres per unit volume (N_v) can be checked from the following, using Table 6:

$$N_v = \frac{1}{A\Delta}[Q_k][n_k]$$

Again the appropriate T/Δ ratio is selected and the number of profiles in each size class is multiplied by the corresponding number from the table and summed to complete the terms $[Q_k][n_k]$.

A complete set of results for pollen tube vesicles is presented in Table 7 using the same set of data as used for the Saltykov area method (Table 4).

Table 5. P_{mk} conversion matrices for size distribution

T/Δ	$m\backslash k$	1	2	3	4	5	6	7	8	9	10
0	1	2·00000	−0·68328	0·08217	−0·02799	−0·00049	−0·00298	−0·00126	−0·00094	−0·00062	−0·00045
	2	0	0·89443	−0·47183	0·04087	−0·02528	−0·00393	−0·00427	−0·00234	−0·00167	−0·00117
	3	0	0	0·66667	−0·39550	0·02903	−0·02416	−0·00516	−0·00488	−0·00288	−0·00208
	4	0	0	0	0·55470	−0·34779	0·02308	−0·02293	−0·00565	−0·00512	−0·00315
	5	0	0	0	0	0·48507	−0·31409	0·01947	−0·02176	−0·00582	−0·00518
	6	0	0	0	0	0	0·43644	−0·28863	0·01704	−0·02071	−0·00584
	7	0	0	0	0	0	0	0·40000	−0·26850	0·01527	−0·01977
	8	0	0	0	0	0	0	0	0·37139	−0·25208	0·01393
	9	0	0	0	0	0	0	0	0	0·34815	−0·23834
	10	0	0	0	0	0	0	0	0	0	0·32880
$\frac{1}{2}$	1	1·00000	−0·23607	−0·01096	−0·00887	−0·00371	−0·00211	−0·00128	−0·00084	−0·00058	−0·00041
	2	0	0·61803	−0·24452	−0·01575	−0·01340	−0·00623	−0·00378	−0·00241	−0·00164	−0·00117
	3	0	0	0·50000	−0·23222	−0·01497	−0·01460	−0·00708	−0·00447	−0·00295	−0·00206
	4	0	0	0	0·43426	−0·21913	−0·01341	−0·01479	−0·00732	−0·00476	−0·00321
	5	0	0	0	0	0·39039	−0·20750	−0·01189	−0·01463	−0·00731	−0·00486
	6	0	0	0	0	0	0·35826	−0·19744	−0·01056	−0·01433	−0·00720
	7	0	0	0	0	0	0	0·33333	−0·18871	−0·00942	−0·01398
	8	0	0	0	0	0	0	0	0·31323	−0·18108	−0·00844
	9	0	0	0	0	0	0	0	0	0·29654	−0·17434
	10	0	0	0	0	0	0	0	0	0	0·28238
1	1	0·66667	−0·12023	−0·01760	−0·00732	−0·00352	−0·00197	−0·00121	−0·00079	−0·00055	−0·00039
	2	0	0·47214	−0·14944	−0·02414	−0·01172	−0·00613	−0·00366	−0·00236	−0·00161	−0·00115
	3	0	0	0·40000	−0·15263	−0·02401	−0·01289	−0·00704	−0·00436	−0·00290	−0·00203
	4	0	0	0	0·35679	−0·15063	−0·02260	−0·01310	−0·00732	−0·00466	−0·00316
	5	0	0	0	0	0·32663	−0·14724	−0·02099	−0·01296	−0·00735	−0·00477
	6	0	0	0	0	0	0·30383	−0·14352	−0·01945	−0·01270	−0·00727

(cont. overleaf)

43

Table 5. (cont.)

T/Δ	$m \backslash k$	1	2	3	4	5	6	7	8	9	10
	7	0	0	0	0	0	0	0·28571	−0·13985	−0·01806	−0·01240
	8	0	0	0	0	0	0	0	0·27081	−0·13634	−0·01681
	9	0	0	0	0	0	0	0	0	0·25825	−0·13305
	10	0	0	0	0	0	0	0	0	0	0·24744
2	1	0·40000	−0·04900	−0·01277	−0·00555	−0·00289	−0·00169	−0·00107	−0·00071	−0·00050	−0·00036
	2	0	0·32071	−0·07251	−0·02024	−0·00961	−0·00535	−0·00329	−0·00217	−0·00150	−0·00109
	3	0	0	0·28571	−0·08035	−0·02205	−0·01094	−0·00631	−0·00400	−0·00271	−0·00192
	4	0	0	0	0·26297	−0·08369	−0·02223	−0·01135	−0·00669	−0·00433	−0·00298
	5	0	0	0	0	0·24621	−0·08512	−0·02181	−0·01139	−0·00681	−0·00447
	6	0	0	0	0	0	0·23303	−0·08561	−0·02116	−0·01126	−0·00681
	7	0	0	0	0	0	0	0·22222	−0·08559	−0·02044	−0·01106
	8	0	0	0	0	0	0	0	0·21310	−0·08527	−0·01970
	9	0	0	0	0	0	0	0	0	0·20524	−0·08477
	10	0	0	0	0	0	0	0	0	0	0·19836
4	1	0·22222	−0·01658	−0·00605	−0·00309	−0·00181	−0·00115	−0·00078	−0·00055	−0·00040	−0·00030
	2	0	0·19539	−0·02811	−0·01104	−0·00599	−0·00367	−0·00242	−0·00168	−0·00121	−0·00091
	3	0	0	0·18182	−0·03351	−0·01316	−0·00729	−0·00457	−0·00307	−0·00217	−0·00160
	4	0	0	0	0·17233	−0·03675	−0·01420	−0·00795	−0·00504	−0·00343	−0·00246
	5	0	0	0	0	0·16497	−0·03890	−0·01473	−0·00827	−0·00528	−0·00363
	6	0	0	0	0	0	0·15895	−0·04043	−0·01497	−0·00843	−0·00541
	7	0	0	0	0	0	0	0·15385	−0·04155	−0·01505	−0·00848
	8	0	0	0	0	0	0	0	0·14942	−0·04239	−0·01503
	9	0	0	0	0	0	0	0	0	0·14551	−0·04303
	10	0	0	0	0	0	0	0	0	0	0·14202
6	1	0·15385	−0·00825	−0·00341	−0·00190	−0·00119	−0·00080	−0·00057	−0·00041	−0·00031	−0·00024
	2	0	0·14049	−0·01482	−0·00664	−0·00390	−0·00254	−0·00176	−0·00127	−0·00095	−0·00073
	3	0	0	0·13333	−0·01827	−0·00827	−0·00494	−0·00327	−0·00231	−0·00169	−0·00128

	1	2	3	4	5	6	7	8	9	10
4	0	0	0	0·12816	−0·02055	−0·00923	−0·00555	−0·00371	−0·00264	−0·00195
5	0	0	0	0	0·12405	−0·02220	−0·00984	−0·00593	−0·00399	−0·00285
6	0	0	0	0	0	0·12061	−0·02346	−0·01025	−0·00617	−0·00416
7	0	0	0	0	0	0	0·11765	−0·02446	−0·01052	−0·00633
8	0	0	0	0	0	0	0	0·11504	−0·02528	−0·01071
9	0	0	0	0	0	0	0	0	0·11271	−0·02595
10	0·11765	0	0	0	0	0	0	0	0	0·11060
8 1	0	−0·00493	−0·00217	−0·00127	−0·00083	−0·00058	−0·00042	−0·00032	−0·00025	−0·00019
2	0	0·10967	−0·00913	−0·00439	−0·00271	−0·00183	−0·00131	−0·00098	−0·00075	−0·00059
3	0	0	0·10526	−0·01148	−0·00561	−0·00351	−0·00242	−0·00176	−0·00133	−0·00103
4	0	0	0	0·10201	−0·01310	−0·00639	−0·00403	−0·00280	−0·00205	−0·00156
5	0	0	0	0	0·09939	−0·01433	−0·00693	−0·00438	−0·00305	−0·00225
6	0	0	0	0	0	0·09717	−0·01530	−0·00732	−0·00462	−0·00323
7	0	0	0	0	0	0	0·09524	−0·01610	−0·00761	−0·00480
8	0	0	0	0	0	0	0	0·09352	−0·01677	−0·00784
9	0	0	0	0	0	0	0	0	0·09198	−0·01734
10	0·09524	0	0	0	0	0	0	0	0	0·09057
10 1	0	−0·00327	−0·00150	−0·00091	−0·00061	−0·00044	−0·00033	−0·00025	−0·00020	−0·00016
2	0	0·08994	−0·00619	−0·00311	−0·00198	−0·00137	−0·00101	−0·00077	−0·00060	−0·00048
3	0	0	0·08696	−0·00788	−0·00404	−0·00261	−0·00185	−0·00137	−0·00106	−0·00083
4	0	0	0	0·08473	−0·00908	−0·00466	−0·00304	−0·00216	−0·00162	−0·00125
5	0	0	0	0	0·08291	−0·01001	−0·00511	−0·00334	−0·00239	−0·00180
6	0	0	0	0	0	0·08136	−0·01076	−0·00545	−0·00356	−0·00255
7	0	0	0	0	0	0	0·08000	−0·01139	−0·00573	−0·00373
8	0	0	0	0	0	0	0	0·07879	−0·01193	−0·00594
9	0	0	0	0	0	0	0	0	0·07769	−0·01240
10	0	0	0	0	0	0	0	0	0	0·07668

Table 6. Q_k, conversion matrices for population densities

T/Δ	1	2	3	4	5	6	7	8	9	10
0	2·00000	0·21115	0·27699	0·17208	0·16331	0·11435	0·09722	0·08436	0·07453	0·06674
$\frac{1}{2}$	1·00000	0·38197	0·24452	0·17741	0·13917	0·11440	0·09707	0·08429	0·07447	0·06670
1	0·66667	0·35191	0·23296	0·17280	0·13674	0·11299	0·09620	0·08368	0·07407	0·06641
2	0·40000	0·27171	0·20043	0·15682	0·12797	0·10770	0·09278	0·08138	0·07242	0·06519
4	0·22222	0·17880	0·14766	0·12468	0·10726	0·09374	0·08297	0·07428	0·06713	0·06117
6	0·15385	0·13223	0·11510	0·10134	0·09014	0·08090	0·07319	0·06669	0·06115	0·05639
8	0·11765	0·10474	0·09395	0·08486	0·07714	0·07053	0·06483	0·05987	0·05375	0·05174
10	0·09524	0·08667	0·07926	0·07283	0·06721	0·06227	0·05791	0·05405	0·05060	0·04752

46

Comparison of the results from each method confirms that the Saltykov method overestimates the vesicle density when applied to thin sections.

It is not always possible to select class intervals (Δ) to give T/Δ ratios that can be used with ten class sizes or less as in Table 5. In these cases new tables should be constructed from the general formulae given in Rose (1980). There is a printing error in the definition of økm, which should read

$$\frac{T}{\Delta} + (m - \tfrac{1}{2}) \left(1 - \frac{(k-1)^2}{(m - \tfrac{1}{2})^2}\right)^{\frac{1}{2}}$$

Sphere size and numerical density from surface–volume relationships. The second general method for estimating sphere size and number depends on the surface area to volume relationships of a sphere, which can be determined from the point count and line intersect counts of the tissue. Since the surface to volume ratio of a sphere is fixed ($4\pi r^2/(4/3)\pi r^3 = 3/r$), then the radius can be calculated from a surface density estimate of the bounding membrane and the volume fraction of the contents ($3/r$ = surface density / volume fraction). Again, for reasons already discussed, this method will only be applicable to populations of spheres with similar sizes, such as nuclei; the margins of error are too large to allow its use for a variable population, such as vesicles.

Table 7. *Vesicle sizes and frequencies*

Class number	Diameter of sections (nm)	Number of sections per μm^2	Mean class diameter of vesicles (nm)	Number of vesicles per μm^3
10	350–316	0·023	332·5	0·132
9	315–281	0·046	297·5	0·216
8	280–246	0·132	262·5	0·674
7	245–211	0·340	227·5	1·804
6	210–176	0·766	192·5	4·166
5	175–141	1·864	157·5	10·982
4	140–106	1·818	122·5	8·570
3	105–71	1.439	87·5	6·076
2	70–36	0·093	52·5	0
1	35–0	0	17·5	0
			Total	32·61

Tradescantia pollen tube vesicles after cytochalasin D treatment. Data from Table 4 were used to calculate vesicle sizes taking into account the section thickness of about 70 nm (courtesy of J. Picton).

The number of spheres per unit volume (N_V) can be calculated from the volume fraction (V_V) information and the numerical density (N_A) of the structures on the micrographs,

$$N_V = \frac{N_A^{3/2}}{V_V^{1/2}} \times \frac{1}{\beta}$$

where β is a shape coefficient that has a value of approximately 1·38 for spheres (Weibel and Bolender, 1973). This formula also assumes that the spheres are all of the same size, however, the correction factor increases this by less than 10% for sphere distributions with standard deviations out to 25% of the mean.

Non-spherical components

Methods for estimating the size and shape of non-spherical particles are necessarily more complex than for spheres. In the first place the three dimensional shape must be established to confirm that it conforms to one of the regular shapes: cubic, cylindrical or ellipsoidal. More complex shapes cannot be handled by the following methods and will not be considered further, apart from pointing out that it would still be worth finding their surface area to volume ratios and using these in comparative studies as indicators of shape and size changes.

Surface to volume ratios can be used to determine basic parameters of cubes (s/v = 6/l, where l is the edge length) and cylinders (s/v = 2/r, where r is the radius), but these can often be obtained by direct measurement in suitable section planes.

Numerical densities can be calculated in the same way as for spheres (p. 47), but using different shape coefficients. The β value for a cubic shape is 1·84, while the values for cylinders and ellipsoids depend on first establishing the ratio of the lengths of their major and minor axes, so that the appropriate value of β can be determined from the standard curves presented by Weibel (1969) and reproduced in Fig. 15.

Experimental design and analysis

The effectiveness of these stereological methods is governed by the design of the analytical system, that is, the selection of the most suitable combination of micrographs with test grids. An effective experimental design depends on a thorough understanding of the three-dimensional organisation of the tissue and cells under consideration, so that sections may be taken in appropriate planes and at appropriate levels in the tissue. Where it is planned to correlate the stereological results with physiological or biochemical observations of the tissue, care must be taken to fix the cells for electron microscopy in a corresponding state so that valid comparisons can be made. Since the total investment, in time, from fixing the tissue to obtaining a complete set of data is quite heavy (several weeks to several months) it is important to be clear from the start exactly what problem is being tackled and what answers are being sought. The following sections discuss each aspect of the operation in turn, from taking the micrographs to establishing the significance of the final results.

Collecting micrographs

The entire analysis depends on the starting material, the micrographs, and no amount of sophisticated stereological and statistical analysis will compensate for an inadequate micrograph collection. Decisions have to be made about the final print magnification of the micrographs and the system to be used for taking an unbiased set from the sections on a grid. These topics are discussed here, while the related question of the number of micrographs required is discussed later (p. 62).

(a) Micrograph magnification

The problems involved stem from requiring high magnification micrographs (final print at $30\,000$–$80\,000\times$) to enable a detailed analysis to be made of a

Fig. 15. Values of the shape coefficient, β, plotted against λ, the ratio of the lengths of the particle axes, for ellipsoids (a) and cylinders (b). For oblate ellipsoids $\lambda < 1$, and prolate ellipsoids $\lambda > 1$ (redrawn from Weibel, 1969).

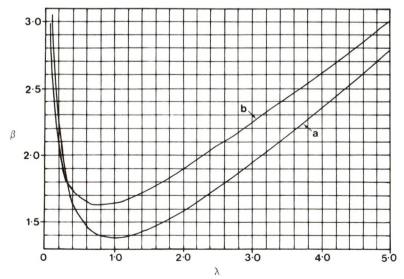

specific cell type, whilst at the same time collecting an unbiased (random) set of micrographs from tissue sections composed largely of structures that are not of direct interest, such as extracellular spaces, other cell types and even nuclei of the cell type under consideration. This problem is especially severe in mature plant cells where more than 90% of the cell volume may be taken up by a large central vacuole, with the cytoplasm limited to a thin peripheral layer.

In general these problems are dealt with by splitting up the analysis into a series of stages, or levels, using micrographs at increasingly greater magnifications between one level and the next. In this way the large-scale features can be handled at low magnifications and the main components of interest examined efficiently at high magnifications. For example, for a tissue made up of two cell types and intercellular spaces, the first level of analysis might be at the light microscope level (using the thinnest possible sections, $0\cdot5$–$1\cdot0$ μm, see p. 71) with photomicrographs at 500–2000×. These are used to establish volume fractions of each cell type in the whole tissue and surface densities of their cell and nuclear boundaries. Further, large-scale features, vacuoles for example, in the cytoplasm of the cell types of interest can be determined at this level. If this is not possible at the light microscope level it can be carried out at the electron microscope level using micrographs at a final print magnification of 3000–7000×. These are obtained by taking sets of micrographs only over the cell type of interest. The contribution of individual cell components to the whole cytoplasm is determined at a third level, taking micrographs only over the cytoplasm of the required cell type, ignoring nuclei and vacuoles, at a magnification of 10 000–20 000×. Finally a fourth level, concerned with, say, the distribution of components within a particular organelle, such as internal membranes, would use micrographs of the organelle at 30 000–80 000×.

This splitting into different levels may appear cumbersome, but in fact it is far more efficient than attempting a single analysis for all components. The number of micrographs required to sample a tissue of the type described above at a high print magnification ($\times 50\,000$) would be very large. Notice that at each level the micrographs are taken over *selected* areas, that is they are not random with respect to the tissue (apart from the first level), but they fulfil the requirement of providing *unbiased* information on components within the selected area.

Information from the highest level (greatest magnification) can be related to the whole tissue by using estimates obtained at successively lower levels for increasing scales of structure. So areas of membrane per unit volume of cytoplasm could be converted to area per unit volume of cell at the next lower level, using the volume fraction of cytoplasm in the cell, and to area per unit volume of tissue using the volume fraction of the cells in the whole tissue. In this way estimates of the total volume, surface area and perhaps number of particular cell components in a whole tissue can be reliably estimated and compared with the biochemical or physiological activities of that tissue.

(b) Collecting unbiased micrographs

The previous section described a method for systematically dividing the analysis between different levels of micrographs from the tissue. Here we will be concerned with the preparation of sections and the collection of unbiased sets of micrographs at each level.

Preliminary examination of the tissue is essential to establish the basic structural organisation of the component cells and their cytoplasmic contents. If these are arranged uniformly within the tissue, such as parenchyma cells in liver, then no particular precautions would be observed about the orientation of the plane of sectioning. Asymmetric distributions, such as cells lying with their long axes parallel to one another, or the endoplasmic reticulum occurring in regular parallel layers in the cytoplasm, require that precautions should be taken to ensure that the section planes are made at a number of angles to the major axis of the tissue. With most tissues there will be some degree of gradation from the outside to the centre, so that section faces from these various depths within the tissue should be prepared.

Normally a given area from each section should only be recorded once, and only one section recorded for each block face. Adjacent sections would not be independent samples of the tissue. Variations between blocks and between tissues sampled from different individuals may also have to be assessed.

Micrographs are collected from suitable sections at magnifications decided by the design of the experiment (p. 49). Each set is taken at a standard magnification on the microscope and lens hysteresis is minimised by always setting the magnification control to the highest level and then returning it to the standard magnification for focussing and exposing the plate. The part of the section photographed should be chosen at random with respect to the content of that area. Some schemes suggest using grid squares as convenient random markers on the sections, and then consistently selecting the micrograph area with respect to the grid bars (e.g., always in the top right-hand sector). However, this method is only suitable for the lowest level of analysis where the whole tissue is being sampled (and this is best done at the light microscope level).

Collecting unbiased sets from one cell type, where the relative contributions of cytoplasm and nucleus to the whole cell volume are to be determined, poses difficulties because the nuclei are so large that they dominate the section and are readily seen with the viewing binoculars. One method of avoiding bias in this situation is to record systematically every single profile of that cell type in the section. Alternatively, low-magnification electron micrographs can be taken and prints made of all areas that include cells of the specified type at the appropriate magnification.

For 'within cytoplasm' micrographs a similar approach can be adopted. Here there is less risk of 'observer bias' as the structures under consideration cannot be seen clearly on the fluorescent screen. Where possible micrographs restricted to the cytoplasm of the required cells are taken, but these may often include 'unwanted areas', such as nuclei, vacuoles and cell walls.

Test grids

Test grids for overlaying micrographs are prepared by drawing the desired pattern in black ink on white paper using narrow lines and discrete points. These can then be used as 'masters' for making transparencies on any office machine used for overhead projector transparencies, or photographically on large film sheets. Distances between points, line length etc. should be carefully checked on the final grid as copying machines do not always produce distortion free copies. Other systems for superimposing the grid on the micrograph have been described, such as printing the negative with a suitable mask so that the grid appears on the final print or by having a series of permanent transparent grids for the screen of a back projection unit on which images of the micrographs are projected from original negatives or film positives.

Design of the grid pattern is a topic which seems to cause endless confusion. It may even be helpful to state clearly at this stage that it is not critical, since it has been demonstrated for both point and line intersect counts that both random and regular arrangements of the test grid design are satisfactory (Weibel, 1969; Weibel and Bolender, 1973). The following paragraphs attempt to take the subject a little further.

(a) Point counting grids

The basic method (p. 29) depends on counting points lying over each type of component and relating these counts to the total count for the area within which the components are contained. In designing the grid the main consideration is the spacing between points on a regular square lattice. Placing the points closer together will increase the number of points over each component, and increase the number of points that have to be counted for a given area of micrograph. It has been stated that exceeding a point count of about one per component profile is a waste of time since there is no further gain in information and more points must be counted. This type of statement neglects the realities of the situation, where the investigator is normally interested in recording the presence of several different components with size ranges of $\frac{1}{4}$–10 μm diameter. The increased accuracy of assessing individual micrographs using a high point density has to be balanced against the additional work and the usually large variation from one micrograph to the next.

A lattice of 100–400 points will usually be suitable for a micrograph on a standard 8 × 10 inch (203 × 254 mm) sheet of photographic paper. Where micrographs contain only a very limited number of profiles of the component under investigation it is possible to increase the 'scoring rate' by using a denser lattice over this component every time that it occurs. This can be accomplished in one of two ways. A standard grid with a large lattice spacing is used

Fig. 16. Square lattice grid, with every fifth line in each direction heavily ruled, giving a ratio of 1 major to 25 minor points. On this grid there are 25 major and 441 minor points. Two component classes are present, A and B. The volume fraction of A is 8 major points divided by 25, or 0·32; and of B is 7/441, or 0·016.

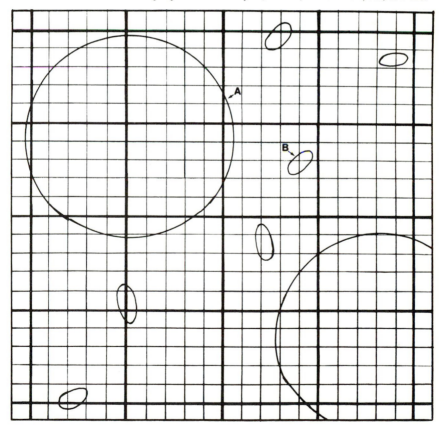

and then a further small grid is placed on top (Bolender, 1978). The extra grid has a lattice with points at, for example, one fifth of the spacing of the main lattice, and is placed in register with the main lattice over the component. The number of these finer points is recorded within the boundary of the component. In this way all the components of this type are recorded for each micrograph; note that the total point count for the test grid will be twenty-five times the number of major lattice points in this example (Fig. 16). Larger and more numerous components are recorded only with the major lattice. Alternatively, a grid of closely spaced lattice points can be prepared and, for example, every fifth row of points in each direction picked out by ruling the line heavily or using a pale coloured marker (Weibel, 1969). Intersects of the heavy or coloured lines would then form the major lattice points. The grids are referred to as square double lattice grids and are characterised by the ratio of

Fig. 17. Estimation of membrane surface area. A regular array of nine parallel lines, each 9 cm long and 1 cm apart, a total line length of 81 cm. The lines intersect the boundaries of three classes of components A, B and C. The total number of intersections with the boundary of A is 10, component B is 12 and component C is 16. The surface density estimate for the A component is $2 \times 10/81 = 0.25$ cm^2/cm^3, for B is $2 \times 12/81 = 0.30$ cm^2/cm^3 and for C is $2 \times 16/81 = 0.40$ cm^2/cm^3.

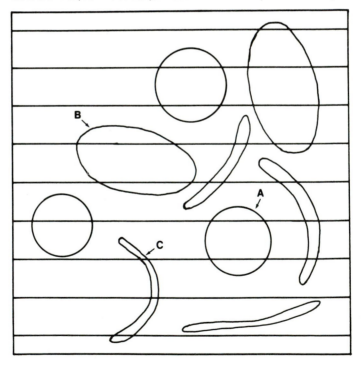

major (coarse) points to minor (fine) points; in the example above it would be 1 : 25.

(b) Line intersect grids

Similar arguments to those stated above apply to the spacing of test lines for surface density determinations. Closely spaced lines will lead to a direct increase in the amount of work involved in scanning along them and only a small increase in the resultant information.

Grids can be devised using intersecting sets of vertical and horizontal lines forming a square grid, that can be used for point counting, with the two sets of lines being used for surface density estimates. Weibel and Bolender (1973) point out that this system leads to an excessive number of intersection counts being recorded, which are not independent since a single component will record at least four counts, although these grids are still extensively used (pp. 94, 102, 113). The number of intersections per component can be effectively reduced by using a single set of parallel lines (Fig. 17). A total of 10–30 lines parallel to the short axis of an 8 × 10 inch sheet are usually adequate, again for particularly rare components a local finer grid can be used. In all cases the total line length within the test grid must be known.

Some biological membrane systems are highly organised forming multi-layered sheets lying parallel to each other, the endoplasmic reticulum of secretory cells, for example. Clearly the number of intersect counts will be maximal where the test grid lines cross such membrane arrays at right angles and minimal where the test lines are parallel to the membranes. For an individual micrograph this does not satisfy the condition for the relationship between intersect counts and surface density (p. 30).

There are several ways of dealing with organised, anisotropic, structures of this type. A large unbiased set of micrographs should include a complete range of membrane orientations with respect to the sectioning plane, and hence the test lines, and if this is so, no further precautions are required. However, these conditions may only be fulfilled by obtaining and scanning an unusually large number of micrographs. This can be avoided by using one of three modifications to the counting system (Weibel, 1969).

The simplest method is to count each micrograph three times, rotating the test grid by 120° with respect to the micrograph between each count. The other two methods rely on using test grids of different designs. In one the test lines are arranged as a series of short lines lying randomly on the test grid and in the other wavy test lines are used. In this latter case the 'waves' are usually built up from semicircular lines to simplify the determination of total test line length on the grid.

(c) Multipurpose test grids

Frequently the investigator is interested in both the volume proportion and surface area of various components in the cytoplasm. This has led to the development of test grids that combine both point and line intersect counting in one operation using one grid. The simplest multipurpose grid is made by joining alternate pairs of dots on one axis of a point test grid with a straight line. Dots on adjacent rows are joined out of register with the rows on either side. This gives a test grid covered in short lines. The ends of the lines correspond to the points of a point lattice, while the lines themselves form a test grid for intersect counts (Fig. 18). Thus a line termination inside a component would score one point, while the line crossing one membrane

Fig. 18. Estimation of volume fraction and surface area. The grid consists of a staggered array of 1-cm lines based on a 1-cm-square lattice. The total number of lines is 36, giving 72 line ends or points, and the total line length is 36 cm. Three classes of component, A, B and C are present. The total volume fraction occupied by component A is 0·125 (9/72), B is 0·139 (10/72) and C is 0·042 (3/72). The surface density estimate of the surfaces of A is 0·28 cm^2/cm^3 (2 × 5/36), B is 0·44 cm^2/cm^3 (2 × 8/36) and C is 0·61 cm^2/cm^3 (2 × 11/36).

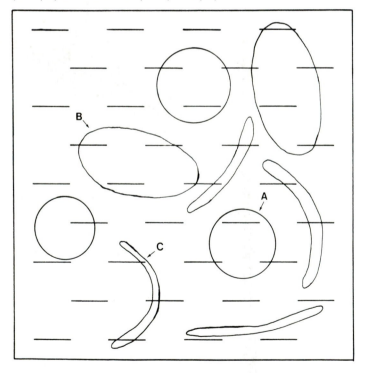

would score one intersection. The total number of points on the grid is twice the number of lines and the total test line length is the number of test lines multiplied by the length of one of the lines.

This simple multipurpose lattice is perfectly adequate, but it can be improved by constructing a system in which all the ends of the lines are the same distance apart within and between the rows (Weibel, 1969). Essentially the grid can be envisaged as a nested set of equilateral triangles, with the bases of alternate triangles along each row drawn in to form the test line (Fig. 19). The intersections of all the triangles are all the same distance apart and these form the test points. Unfortunately this system is slightly more difficult to construct and it is less easy to work out the area of the micrograph being sampled, however these comments should not deter the investigator from using it. Isometric graph paper can be used to prepare the master sheet for making transparent overlays, on an appropriate scale.

Counting and recording

This section will attempt to anticipate, and answer, the numerous questions that arise during the counting process in a stereological analysis. Starting with the first micrograph, it is overlaid with the selected transparent test grid and the two are clamped firmly together on a clipboard or with paper clips so that they do not move relative to one another during counting. The test grid is scanned line by line and individual point or intersect counts recorded. Recording can be done simply by making ticks on a piece of paper, but this is very tedious for large numbers of micrographs. For a limited number of components hand tally counters can be used; these can be secured to a base board for larger analyses. A slightly more sophisticated system is to make use of one of a number of commercially available counters for blood cell counting. These usually have sufficient capacity, about ten channels, for a fairly comprehensive analysis of components with a multipurpose test lattice, recording both point and intersect counts. Totals should be obtained and kept for each micrograph individually, as this information will be required later for statistical analysis.

Problems arising in point counting are concerned with deciding whether a point is inside or outside the boundary of the component of interest. Those occurring exactly over the bounding membrane should be assigned on the alternate 'one in – one out' rule. If the boundary is quite thick then a decision must be made about whether the boundary (e.g. plant cell wall) is to be included or excluded from the component volume and points counted to the outside or inside edge of the boundary accordingly.

Line intersect counts are straightforward when the membrane lies perpendicular to the sectioning plane. Membranes sloping in the thickness of the section cause problems because the relationship between intersect counts and surface density depends on applying the test grid to a true two-dimensional specimen. Steeply sloping membranes can be recorded as a single intersect, they are usually of limited area on the micrograph.

Tangential sections of membrane sheets can occupy quite large areas, and intersect with several test lines. These membranes can be envisaged as saucer-

Fig. 19. Estimation of volume fraction and surface area. A series of 1-cm lines is formed from rhombi, side 1 cm. The vertical distance between lines is half the line length times $\sqrt{3}$, the total grid width is 7 cm (4 lines and 3 spaces) plus the space between the end of the line and the edge of the grid, 0·25 and 0·75 cm, a total width of 8 cm. The total height of the grid is 10 times the interline spacing (9 spaces plus a half space at top and bottom).

The total number of lines is 40, giving 80 line ends or points, and the total line length is 40 cm. Three classes of component, A, B and C are present. The total volume fraction of A is 0·125 (10/80), B is 0·125 (10/80) and C is 0·063 (5/80). The surface density estimate of the surfaces of A is 0·30 cm^2/cm^3 (2 × 6/40), B is 0·30 cm^2/cm^3 (2 × 6/40) and C is 0·55 cm^2/cm^3 (2 × 11/40).

shaped structures lying in the depth of the section, with the rim of the saucer exposed at one section face. The section face is a true two-dimensional plane, and a test line in that plane will intersect with the saucer rim at two places, registering two counts. A test line on the opposite face of the section will not intersect with the saucer at all, so no counts are registered. Unfortunately it is not possible to distinguish between section faces in an electron micrograph, however it may be assumed that in an unbiased collection of micrographs the 'saucer' will occur with equal frequency at each section face. So the problem can be resolved by recording one intersect count for each test line that crosses a tangentially sectioned membrane profile (one count is an average of two counts and zero counts for each section face).

The total number of points and length of test lines (adjusted for micrograph magnification) must be recorded for each micrograph sampled. Where the area of interest covers the whole micrograph, these will correspond to the totals for the test grid. However in many cases unwanted areas intrude on the micrograph, for example nuclei or extracellular space on micrographs concerned with estimating components on a unit volume of cytoplasm basis. In these cases the total point count and line length lying over the desired area alone must be recorded. This is simple for point counts, but for line intersects the line length must be measured when continuous lines are being used. Where a multipurpose grid of short lines is being used, only counts from complete lines are recorded, those lying across the boundary with the unwanted area are ignored. It should be noted that recording counts from less than the whole test grid may introduce complications with the statistical analysis.

Statistical analysis
While averaging results from a number of micrographs gives an indication of the magnitude of the volume or surface area values for the system under study, it is desirable that the appropriate statistical analysis be carried out before any firm conclusions are reached. This is especially true where comparisons are being made between different developmental stages or different experimental treatments. A consideration of the requirements for statistical analysis before commencing any work will often save much time and duplication of effort later. For a large-scale analysis it is often worth while running a preliminary experiment to determine where the main sources of variation are likely to occur. This should produce information on the number of independent specimens that need to be fixed and embedded, the number of blocks that need to be examined from each, and their orientation, and the number of micrographs required from each section.

Means and confidence limits

Each micrograph is regarded as a single observation and the test grid counts are used to determine a single value for each parameter (volume fraction, surface density, etc.) for that micrograph. These are used to determine the mean, standard error and standard deviation for each group of micrographs using standard statistical methods assuming a normal distribution (see, for example, Bailey, 1959; Snedecor and Cochran, 1967; Sokal and Rohlf, 1969; Parker, 1979). If the mean count levels are very low the individual parameter values for the group of micrographs may conform to a Poisson distribution. This can be tested by examining the kurtosis and skewness of the distribution. Confidence limits for such skewed distributions can be found either by transforming the data to a normal distribution (for example, by taking the square root of each value) or by using the appropriate formula for calculating these limits directly (see, for example, Parker, 1979, p. 33). The mean values for the observations from each block will be normally distributed, and the information from each block can be combined to give an estimate of each parameter for each organelle in the tissue and whole organism.

Treatment comparisons

Here 'treatment' will be considered to mean either organisms exposed to different conditions, or to organisms allowed to progress along a developmental pathway for different periods of time (as when following a cell differentiation sequence, for example).

Tests for significant differences between organisms from different treatments are made using a hierarchic analysis of variance (Sokal and Rohlf, 1969, chapter 10; Parker, 1979, chapter 10.5). A typical hierarchy might be treatments, rats, livers, blocks and micrographs. Notice that the analysis is correctly performed by comparing data for *animals* in this case, that is at the highest level in the table. The information from the lower levels is used to determine a single mean or total value for each animal (e.g. surface density of endoplasmic reticulum in liver, or total number of mitochondria in liver). The values for each animal in each treatment are then compared, computing the mean square error at this level. Notice that no valid comparison between treatments can be made unless there is replication at this highest level, regardless of the level of replication elsewhere, because any differences detected may be due to chance variation between the individuals selected (see Parker, 1979, problem 10.5).

In some cases the variance levels of the observations may not be homogeneous between treatments. This might occur when the treatment resulted in large changes in volume fraction or surface density of components, leading to a large change in the number of counts for components recorded per micrograph. While this could be offset by changing the number of micrographs (replicates), it would lead to undesirable complications in the analysis of variance. Several methods are available for coping with this problem; these are detailed in Sokal and Rohlf (1969, chapter 13).

In quantitative microscopy the number of individual animals or plants examined is usually very low (one to five), causing problems in the reliable estimation of error at this uppermost level of replication because the mean squares are based on only a few degrees of freedom. A more reliable estimate can be made using the mean square from the next lowest level of replication which has a larger number of degrees of freedom.

The complete analysis of variance table also provides useful information on the design of the experiment enabling improvements to be made to the design of future experiments of a similar nature. Those levels in the hierarchy which have highly significant F ratios probably contain additional variance and future work should attempt to introduce a greater degree of replication at these levels. The most severe problems are usually encountered at the lowest level, the micrographs. If the F ratios at this level are very low, or less than 1.0, the implication is that the number of micrographs should be increased

Table 8. *Membrane content of expanding leaves of* Cucumis sativus

Membrane system	Surface area factor increase	Relative area increase compared with plasma membrane
Plasma membrane	31·71	1·00
Tonoplast	41·24	1·30
Golgi apparatus and vesicles	31·29	0·99
Endoplasmic reticulum and nuclear envelope	7·45	0·23
Outer mitochondrial membrane	9·80	0·31
Inner mitochondrial membrane	12·42	0·39
Choloroplast envelope	39·95	1·26
Chloroplast lamellae	48·97	1·54

The actual and relative factor increases in the area of the membrane systems during a 94-fold increase in leaf volume. There has been a relative decrease in the area of endoplasmic reticulum and mitochondrial membranes, and an increase in the area of chloroplast membranes and tonoplast. The Golgi membranes remain unchanged (data from Forde and Steer, 1976).

in future work. This topic will be considered further in the next section.

Evaluation of treatment effects on a particular cell component are made initially on a unit volume of cytoplasm basis. Conversion of this data to a whole cell basis may not be easy, and probably undesirable since neither unit volume of cytoplasm or whole cell volume may be the best reference points, especially where the treatment results in changes in the volume of the cell or cytoplasm. A better idea of the treatment effects can be obtained by comparing levels of the affected component in the cell with the levels of another, unaffected, component (Bolender, 1978), if possible one of similar size, shape and distribution. It is preferable to select the parameter of the reference component that is least likely to change during the treatment, for example, the surface density of the bounding membrane. All the control data are set to 100% (or unity) and the relative changes from control to treatment values for each are multiplied by the ratio of the control : treatment values for the reference parameter (Table 8).

Sample size

The number of observations (micrographs) required depends on the level of accuracy required. It may not be possible, even with a very large-scale analysis, to reduce the level of uncertainty to a low level in this type of work. In many cases this is not necessary, for example, when there are large-scale differences between treatments or when only an approximate value is required for each component.

Clearly homogeneous cytoplasm containing large numbers of evenly dispersed components can be adequately sampled by relatively few micrographs, while rare components or highly asymmetric cells, for example, those with large vacuoles or polarised secretory cells, will require a much higher number of micrographs.

The number of micrographs required for a given level of accuracy in a particular analysis can only be predicted after a preliminary sample has been examined. The observations from between 10 and 20 micrographs, collected by the same method that it is intended to use throughout the analysis, are used to find the mean ($\bar{x}$) and standard error (s) of this small sample. The number of micrographs (n) can then be estimated (Weibel, 1969) for the desired 95% confidence limit percentage of the mean (y):

$$n = (200s/y\bar{x})^2$$

A progressive method has been suggested by Bolender (1978) in which the trial sample is split up into groups and the mean and standard error computed for successively larger numbers of micrographs. The decline in the standard error level with increasing number of observations is used to calculate the approximate number required for a particular level of the standard error. An example of the effect of increasing the number of micrographs and the number of organisms is illustrated in Fig. 20.

The number of micrographs required for point counting determinations can be found for a given level of accuracy, without a preliminary experiment, by consulting the nomogram from Weibel and Bolender (1973), reproduced in Fig. 21. In general, greater efficiency is achieved by increasing the number of micrographs sampled, not the number of points from each micrograph (Nicholson, 1978).

The general problem of the amount of replication required at each level of the analysis falls into the area of statistics known as 'optimal allocation of resources' (Sokal and Rohlf, 1969, p. 289). The methods enable an assessment to be made of the relative amount of effort (time and cost) required at each level, that is, the number of organisms, blocks and micrographs required (examples are given in Sokal and Rohlf, 1969, Box 10.6; Parker, 1979, chapter 10.5).

Frequently the experimental design cannot be completed before such a trial experiment and assessment has been completed. This assessment may indicate

Fig. 20. Surface density of rough endoplasmic reticulum from rat pancreatic exocrine cells: the effects of inadequate replication. The mean (100%) and standard error ($\pm$ 6%) of surface density estimates from 60 micrographs for each of five animals from a uniform population are represented by the horizontal solid and dashed line respectively. The effect of reducing the initial sample size from five animals to two is illustrated by taking data from pairs of animals (ten combinations) and calculating means and standard errors for (a) 20 micrographs per animal and (b) 60 micrographs per animal. From limited data it would be concluded that certain pairs of animals contained significantly different levels of endoplasmic reticulum from others, and this situation is not improved by increasing the number of micrographs examined from each animal (redrawn from Bolender, 1978).

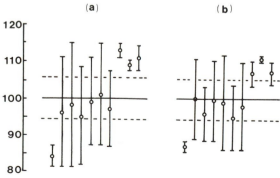

that a large number of micrographs or organisms are required for each treatment, indicating that the experimental design should limit the number of treatments so that it is practicable, in terms of either time or financial cost, to complete the experiment.

Data handling

Statistical analysis of results from test grid counting (p. 57) can be handled on any calculator designed for ordinary statistical work (i.e. one having Σx and $\Sigma (x)^2$ facilities). When the final sample size is not known, and intermediate results are required to assess the level of the standard error, there is much repetition involved in entering the same sets of results on consecutive runs. In this case it may be worth considering using a more sophisticated data handling system. Most teaching and research organisations contain statistics units that will compute the desired statistical parameters using standard programmes. These facilities will have memory stores capable of retaining the information from each micrograph, so that it can be used on successive runs. Alternatively a separate punched card can be prepared for the data from each micrograph and these can be grouped together or added to as desired for each analysis.

Fig. 21. Nomogram for estimating the total number of points (P) that need to be counted to achieve a particular level of relative error (E). The family of curves is for components with different volume fractions, from 0·05 (curve a), 0·1, 0·2, 0·4, 0·6, 0·8 and 0·9 (curve b) respectively. (Redrawn from Weibel, 1963.)

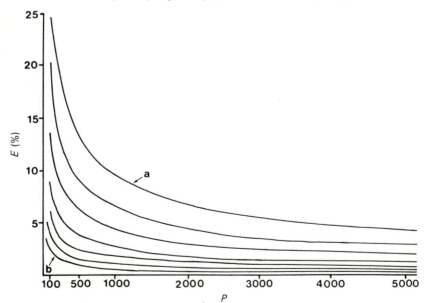

Systems are available for on-line determination of volume and surface density values, and their statistical parameters, during the counting process. Also various companies market sophisticated systems for the entire stereological analysis from the design of the test grid, to counting and presentation of final results. Considering the financial investment involved, it would be worth while gaining experience with the systems described in this book, so that a realistic appraisal of the value of stereology in a particular circumstance can be made before a final decision is reached. Usually it will be found that a great deal of information can be obtained with relatively simple, and inexpensive, methods.

Absolute volume and surface area values
Stereological analysis yields volume and surface area values for components per unit volume of cytoplasm or cell. Whilst they may be useful in this form it is often necessary to convert them to absolute values for whole cells or the whole tissue to enable comparisons with biochemical or physiological data to be made. Whole tissue volumes can be easily determined where a complete organ is involved, such as whole liver or whole leaf blade, but many tissues do not form discrete accessible organs. Some cells have a regular geometric shape, as with spherical or ellipsoidal unicellular organisms, and cylindrical or cuboidal tissue cells, so that total cell volumes can be determined from measurements made with the light microscope on whole cells or on sections made in appropriate planes.

In all of the above cases it is important to establish the method of volume determination at the outset, so that the sample micrographs are unbiased with respect to the whole cell or tissue volume. For example, if the cell size is determined from whole cells, the measurements will be made to the outermost margin of the cell, so stereological analysis should do likewise, even if a cell wall is present. Similarly tissue samples must be unbiased with respect to all the components from within the organ to the outermost margins. Where such an analysis includes unwanted cells or structures, such as blood vessels in liver or veins in leaves, these must be included in the first level of analysis and their contribution to the whole tissue determined. Subsequent levels of analysis can concentrate on the cells of interest (p. 50).

Volumes of irregularly shaped cells can be determined indirectly by finding the volume fraction and absolute volume and number for one of the components. Spherical nuclei form ideal objects for this approach, the number per cell is usually fixed at one, though some tissues are regularly binucleate (e.g. plant anther tapetum cells) and others irregularly binucleate (liver). Determination of nuclear size and volume has been discussed previously (p. 33).

Errors and correction factors

Embedding, sectioning and photography

Ideally the size, shape and distribution of cellular components in the final micrograph should correspond closely to that which was present in the living tissue. Unfortunately considerable changes can be introduced by all the preparation stages from fixation to the final printing. Fixation of the living tissue can result in immediate, and sometimes dramatic, changes in tissue volume. Animal tissues are especially sensitive to the osmolarity of the fixative solutions, swelling or shrinking accordingly. These changes are not necessarily uniform, resulting in shape changes of the tissues and component cells. Such gross changes should be avoided, checking the dimensions of tissue blocks before and after fixation provides a crude but simple method of monitoring these effects.

Differential changes within cells are more difficult to detect, estimate and compensate for. Usually these result from specific interactions between the contents of an organelle and the fixatives and dehydrating agents. Mucopolysaccharide and similar carbohydrate contents of vesicles and vacuoles are particularly susceptible to volume changes, swelling in the presence of ions from the buffer systems and collapsing in dehydration. These changes can be detected subsequently in thin sections. The *in vivo* vesicle shape is more or less spherical, with a smooth membrane. Swelling results in rupture of this membrane while shrinkage results in a membrane that is distorted and wavy in outline.

Rapid dehydration can cause shrinkage of tissues, but more frequently this occurs on transfer from the solvent to the first mixture of solvent and embedding plastic (infiltration). The effect is more severe with the more viscous plastics, such as epon-araldite mixtures, and less so with low-viscosity resins such as Spurr's. These effects appear to result from the solvent leaving the tissue faster than the resin can enter, and are avoided by slow infiltration of the tissue blocks.

Sections are cut on glass or diamond knives, a process which can result in severe distortion due to compression effects as the block face travels past the cutting edge. The extent of this distortion can be estimated by examining the profiles resulting from sectioning spherical structures such as nuclei. The ratio of their diameter parallel to the direction of sectioning to their diameter at right angles to this direction provides a quantitative measure of the compression suffered by the section (Loud, Barany and Pack, 1965). Use of sharp knives and narrow block faces, followed by expansion of the sections while floating on the water bath with solvent vapour should eliminate such problems. Hard resin blocks will clearly cause less trouble than soft ones.

Distortion introduced by the electron optical system of the microscope should be corrected before the micrographs are taken. A diffraction grating replica can be used for calibrating the microscope magnification steps, latex spheres are much less reliable. Reproducibility in using the same magnification step can be achieved by starting from the highest magnification setting on the instrument and coming down in steps to the desired magnification step on each occasion.

The negatives from the microscope are usually on a very stable base, glass or polyester, and do not suffer size changes with humidity as with acetate based films. Photographic printing and glazing can result in changes in the dimensions of the final print after its exposure to the negative. Careful measurement of a few sheets of paper before and after processing will enable these effects to be monitored.

Section thickness

The stereological methods described here were developed for use on polished faces of solid specimens. Electron micrographs are projections of three-dimensional structures, the sections, on to a two-dimensional surface, the micrograph. For structures much larger than the section thickness these projected images closely approximate to a polished surface. However, many cell structures are very small, such as vesicles with diameters down to about 30 nm, and their projected images differ considerably from those expected on ideal two-dimensional surfaces.

Thin sections often include parts of small structures whose membranes curve in the depth of the section, for example, tangential sections through the periphery of vesicles and oblique sections through endoplasmic reticulum sheets. These curving fragments will be projected on to the final two-dimensional micrograph. Two types of error result from this projection, depending on the contrast of the membrane and the staining density of the enclosed compartment. Where the compartment density is low there is an apparent increase in membrane thickness due to its curvature, so decreasing the projected surface area proportion of the enclosed compartment on the micrograph, leading to an underestimate of its volume fraction (Fig. 9). Also if the membrane fragments are very small and of low contrast, they may not be visible against the background of surrounding and overlying cell components, so that both compartment volume and membrane surface area are underestimated.

Where the membrane and enclosed compartment are of high contrast and staining density they are not separable. In this case they appear to be present to the same extent throughout the entire section thickness, and point counting to the edge of the resultant profile will lead to an overestimation of their contribution to the cell (the Holmes effect, Fig. 9).

Clearly the extent and direction of the error depends on the minimum amount of the particle that can be detected in a thin section and on its size relative to the section thickness. Weibel and Paumgartner (1978) have derived expressions for such correction factors for the three regular geometrical shapes (spheres, discs and cylinders) that most closely approximate to the shapes of the cell components. Many nuclei and vesicles approximate to spherical shapes; while mitochondria can, in some cases, be considered as cylinders; and the cisternae of endoplasmic reticulum and of the Golgi apparatus are considered as discs.

The correction factor expressions developed by Weibel and Paumgartner (1978) are too cumbersome to be useful to most cell biologists, but they have also published sets of standard graphs for surface density and volume fraction correction factors for each shape (Figs. 22 to 27). To use these graphs estimates of the section thickness and component dimensions are required.

A crude estimate of the section thickness, probably sufficient for use in this context, can be made by observing the interference colours of the sections after expansion in the microtome knife boat (Table 9). Other methods rely on observation of the sections in the electron microscope. Folds or pleats sometimes occur in the sections and their minimum size corresponds to an edge on view of two section thicknesses appressed face to face, so that half the minimum width corresponds to the section thickness (Small, 1968). Internal standards, either naturally occurring or introduced, can also be used. In general accurate determination of section thickness is difficult (Hayat, 1970; Reid, 1974) and not strictly necessary for this work.

Table 9. *Section thickness from interference colours*

Colour	Thickness (nm)
grey	30–60
silver	60–90
gold	90–150
purple	150–190
blue	190–240

The component dimensions can be estimated to give order of magnitude figures. The smallest dimension (d) of the whole structure, corresponding to the diameter of spheres and cylinders and the thickness of discs, is used with the section thickness, T, to calculate the relative section thickness, g, where

$$g = T/d$$

For spheres the truncation effects can be quite serious, and some estimate of the radius of the smallest cap section detectable is required to calculate the fraction (p) that this radius represents of the sphere radius. Values of p range

Fig. 22. Correction factors: sphere volume fractions. Each of the family of curves gives correction factors for the ratio of the radius of the smallest section profile detectable to the radius of the sphere, from ratio values of 0 (curve a), 0·6, 0·8, 0·9 to 1·0 (curve b) respectively. (Redrawn from Weibel and Paumgartner, 1978.)

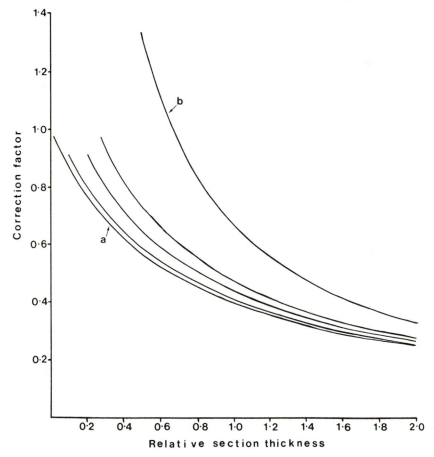

from zero, where even the smallest cap section is detectable, to one, where only median sections of the component can be detected (Figs. 22 and 23).

Discs and cylinders are less likely to be sectioned in planes that produce small truncated fragments in the section thickness. However, these components can be of various proportions, depending on the ratio of their dimensions. For discs this is δ, the ratio of diameter to thickness, and for cylinders it is λ, which is the ratio of cylinder length to diameter. The correc-

Fig. 23. Correction factors: sphere surface densities. Each of the family of curves gives correction factors for the ratio of the radius of the smallest section profile detectable to the radius of the sphere, from ratio values of 0 (curve a), 0·6, 0·8, 0·9 to 1·0 (curve b) respectively. (Redrawn from Weibel and Paumgartner, 1978.)

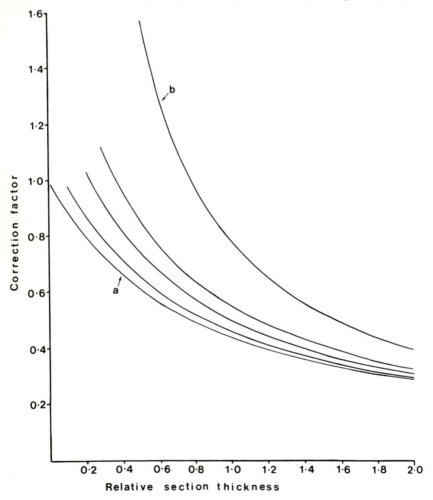

tion factors are relatively insensitive to these ratios (Figs. 24 to 27), so that they need not be determined accurately.

Consideration of these graphs leads to the general conclusion that accuracy will always be improved by cutting thinner sections, even when light microscope observations are being made on large structures such as nuclei. The correction factors are large, even under conditions of contrast and section thickness that would be considered ideal. This emphasises the care with which the counting procedures should be carried out. However the reader should remember that comparative studies of similar systems will yield reliable estimates of differences between the specimens, since the correction factors will be similar in each case.

Fig. 24. Correction factors: cylinder volume fractions. The curves give correction factors for the ratios of cylinder length to diameter of 5 (curve a) and 100 (curve b). (Redrawn from Weibel and Paumgartner, 1978.)

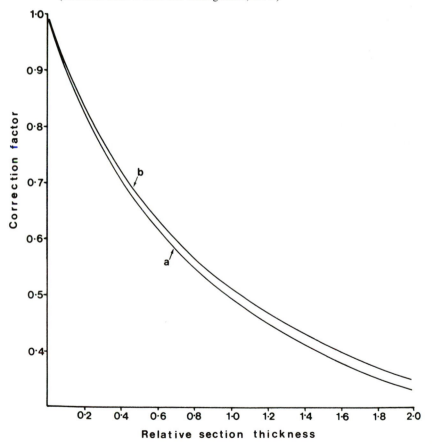

Finally it should be noted that the correction factors for surface density estimates derived by Weibel and Paumgartner (1978) frequently have values less than 1·0 (Figs. 22 to 25), which is in direct conflict with earlier estimates based on those of Loud (1968) who used factors of 1·5 on the grounds that obliquely sectioned membranes are invisible in micrographs. Weibel and Paumgartner (1978) emphasise that their correction factors assume that high contrast micrographs are examined at high magnifications.

Fig. 25. Correction factors: cylinder surface densities. Each of the family of curves gives correction factors for the ratios of cylinder length to diameter from 5 (curve a), 10, 20 to 100 (curve b) respectively. The cylinders are assumed to be open ended. (Redrawn from Weibel and Paumgartner, 1978.)

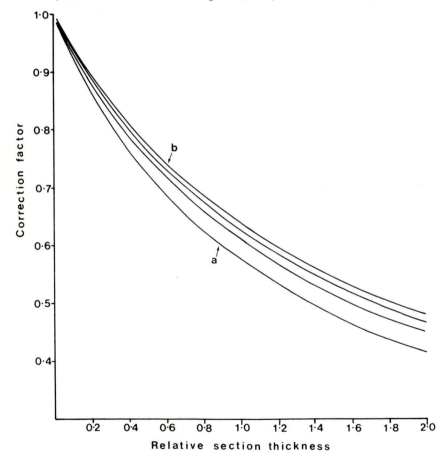

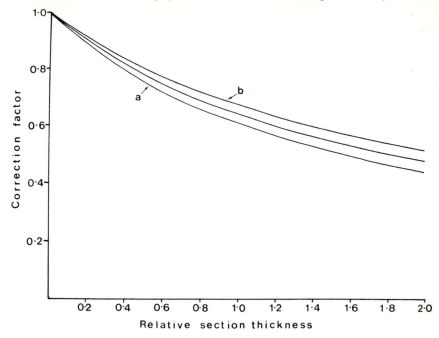

Fig. 26. Correction factors: disc volume fractions. Each of the family of curves gives correction factors for the ratios of disc thickness to diameter from 5 (curve a), 10 to 50 (curve b) respectively. (Redrawn from Weibel and Paumgartner, 1978.)

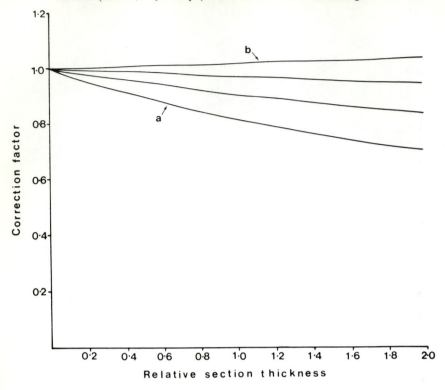

Fig. 27. Correction factors: disc surface densities. Each of the family of curves gives correction factors for the ratios of disc thickness to diameter from 5 (curve a), 10, 20 to 50 (curve b) respectively. (Redrawn from Weibel and Paumgartner, 1978.)

PART IV

Examples and applications

Introduction

Confidence in the use of the quantitative methods discussed in the previous chapters will only be gained by experience. While every effort has been made to set out the theory and practical application of the methods in a readily comprehensible manner, it is recognised that many would like some confirmation that they are applying them correctly. The first part of this chapter provides this by discussing a number of basic examples in detail, presenting as much of the raw data and calculations as possible.

There are many applications for these quantitative methods and this is reflected in the great diversity of the research work that has employed them. Some of them are examined in this chapter to illustrate both the specific approaches adopted and the range of biological problems investigated. Particular emphasis is placed on details of the sampling systems used in preparing the micrographs, the test grids employed, the statistical treatment of the results and the nature of the correction factors used. It will become apparent that most effort has been concentrated on relatively homogeneous cells and tissues which either have clearly defined functions, or exhibit specific responses to particular conditions.

Examples

The following examples have been chosen to illustrate the calculations involved in quantitative microscopy using two readily available tissues, root tips and liver. All the major types of test grid are employed, for the purposes of illustration rather than because the two tissues require different treatments.

Plant root tip

Plant tissues are difficult subjects for the application of quantitative microscopy methods since they usually contain heterogeneous collections of highly
vacuolated cells. However, the cell outlines are readily distinguished by their
enclosing cell walls and limited regions of relatively uniform cells with small
vacuoles can be found, for example, the root tip.

Sampling levels

Two levels of sampling (light micrographs, $\times 1700$; electron micrographs,
$\times 25\,000$) were employed to examine the nuclei, vacuoles, plastids and mitochondria of the cells. The light micrograph was analysed using a square lattice
of lines, 1 cm apart, while for the electron micrograph an interrupted line grid,
based on a square lattice of points with 1 cm spacing was used.

Counts

The light micrograph (Fig. 28) is covered with a total of 423 cm of test line,
2490 μm corrected for micrograph magnification, and 216 line intersections or
points.

> *Points*
>
> | cell walls and extracellular space : | 10 |
> | cytoplasm | : 117 |
> | nuclei | : 33 |
> | vacuoles | : 56 |
> | total | : 216 |
>
> *Line intersections*
>
> | nuclear envelope | : 86 |
> | tonoplast | : 341 |
> | plasma membrane | : 301 |

The line intersect counts are based on the knowledge that plasma membrane lies between the cell wall and cytoplasm, nuclear envelope surrounds
each nucleus and tonoplast separates the vacuoles from the cytoplasm.

The test grid superimposed on the electron micrograph (Fig. 29) contains a
total of 36 μm of test line, corrected for the magnification and 180 line
terminations or points.

> *Points*
>
> | cytoplasm (excluding nucleus and vacuoles) : | 133 |
> | plastids | : 9 |
> | mitochondria | : 13 |
> | nucleus | : 18 |

Fig. 28. Cells of a bean (*Vicia faba*) root tip stained with toluidine blue. Cell walls, nuclei and vacuoles are distinguishable in this light micrograph of a one micron section (×1700). The sampling grid consists of horizontal and vertical lines with 1-cm spacing.

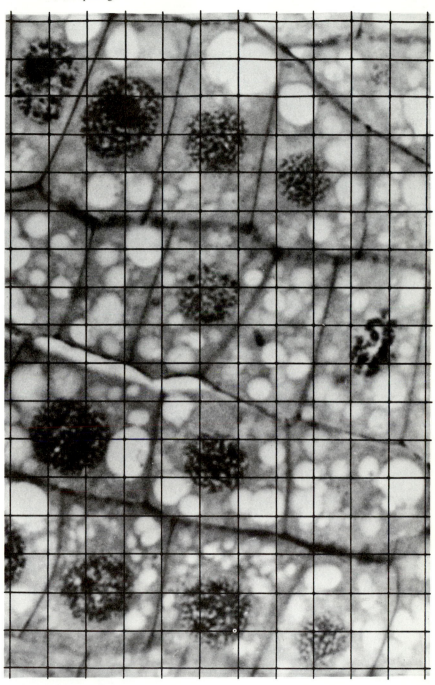

Fig. 29. An electron micrograph of a thin section of bean root tip. Cell walls, vacuoles, mitochondria, plastids and part of a nucleus are among the components present (×25 000). The micrograph is overlaid with an interrupted line lattice based on a 1-cm-square grid.

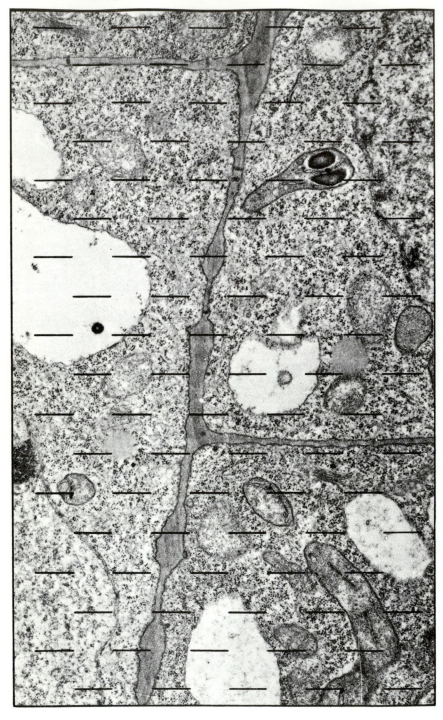

| vacuoles | : 20 |
| total | : 180 |

Line intersections

plastid envelope	: 7
mitochondrial envelope	: 13
nuclear envelope	: 6
tonoplast	: 14
plasma membrane	: 16

The electron micrograph was taken over an area selected for its cytoplasmic content. It is, therefore, not representative of the nuclei, vacuoles or cell walls of the cell so that the counts from these components and their associated membranes are of little value. The sample area of micrograph for the plastids and mitochondria must be restricted to this non-vacuolar part of the cytoplasm.

Calculations: cell size, nucleus, cytoplasm and vacuoles

These may be estimated from the results at level I. The nuclear volume is found from their surface to volume ratio, and the cell size from the volume fraction.

Volume fraction of nucleus in tissue = $33/216 = 0.1528 \ \mu m^3/\mu m^3$
Surface density of nucleus in tissue = $(2 \times 86)/2490$
$$= 0.0691 \ \mu m^2/\mu m^3$$
Average radius of nucleus, $r = (3 \times 0.1528)/0.0691 = 6.63 \ \mu m$

Average volume of nucleus = $1220.8 \ \mu m^3$
Volume fraction of nucleus in cell = $33/206 = 0.1602 \ \mu m^3/\mu m^3$
Average cell volume = $1220.8/0.1602 = 7620.5 \ \mu m^3$

Average cytoplasmic volume = $7620.5 - 1220.8 = 6399.7 \ \mu m^3$
Average vacuole volume per cell = $(56/206) \times 7620.5$
$$= 2071.6 \ \mu m^3$$
Average non-vacuolar cytoplasm per cell = $6399.7 - 2071.6$
$$= 4328.1 \ \mu m^3$$

Plasma membrane and tonoplast

These may be determined from the light micrograph at Level I. The surface density of the membranes can be determined and the absolute level per cell found by one of two methods.

Surface density of plasma membrane in tissue = $(2 \times 301)/2490$
$$= 0.2418 \ \mu m^2/\mu m^3$$
Surface density of tonoplast in tissue = $(2 \times 341)/2490$
$$= 0.2739 \ \mu m^2/\mu m^3$$

Method a. From the intersection counts we know the ratio of the nuclear surface to both the other membranes, the nuclear radius has been determined above.

Average nuclear envelope area per cell = 552·38 μm^2
Plasma membrane area per cell = (301/86) × 552·38 = 1933·3 μm^2
Tonoplast area per cell = (341/86) × 552·38 = 2190·3 μm^2

Method b. The surface density values can be adjusted to give the area per unit volume of cell and then multiplied by the cell volume.

Surface density of plasma membrane in the cells
= 0·2418 × (216/206) = 0·2535 μm^2/μm^3
Area of plasma membrane per cell = 0·2535 × 7620·5
= 1931·8 μm^2
Surface density of tonoplast in the cells
= 0·2739 × (216/206) = 0·2872 μm^2/μm^3
Area of tonoplast per cell = 0·2872 × 7620·5 = 2190·3 μm^2

It should be noted that these methods are merely alternative ways of performing the same determination, they are not independent since they both rely on the initial estimate of the nuclear radius.

Plastids and mitochondria

The total volume of these organelles per cell can be found from the volume fraction of the organelles in the cytoplasm (excluding vacuoles and nucleus) and the average volume of this part of the cytoplasm.

Average total plastid volume per cell = (9/133) × 4328·1
= 292·9 μm^3
Average total mitochondrial volume per cell
= (13/133) × 4328·1 = 423·0 μm^3

Total surface area of mitochondrial and plastid envelopes per cell can be found from the surface density of these membranes in the cytoplasm (excluding vacuoles) and the volume of this part of the cytoplasm in the cell. The total test line length for this calculation can be found either by counting all the complete lines and estimating the extent of the lines partially lying over this part of the cytoplasm, or by multiplying the total line length by the area fraction of the cytoplasm on the whole grid.

The first method estimates a total of 68 lines, each 0·4 μm long, giving a total of 27·2 μm.

In the second method the area fraction is 133/180, the total line length is

36 μm, giving a total of 26·6 μm. In the following this second value will be used.

$$\text{Average plastid envelope surface density} = (2 \times 7)/26\cdot6$$
$$= 0\cdot5263 \ \mu\text{m}^2/\mu\text{m}^3$$
$$\text{Total plastid envelope per cell} = 0\cdot5263 \times 4328\cdot1$$
$$= 2277\cdot9 \ \mu\text{m}^2$$
$$\text{Average mitochondrial envelope surface density}$$
$$= (2 \times 13)/26\cdot6 = 0\cdot9774 \ \mu\text{m}^2/\mu\text{m}^3$$
$$\text{Total mitochondrial envelope per cell} = 0\cdot9774 \times 4328\cdot1$$
$$= 4233\cdot5 \ \mu\text{m}^2$$

Organelle numbers and volumes

The individual plastids exhibit a considerable range of morphology as judged by their profiles in the micrographs. However, the mitochondria appear to be more uniform in shape, approximating to ellipsoids. These have an axial ratio of about 1·5, determined by measuring the mean width and length of the oval profiles. This allows the determination of their numerical density by the shape coefficient method (p. 48).

The number of mitochondrial profiles present on the micrograph is 12, and the total area is $4\cdot48 \times 7\cdot4 \ (= 33\cdot15)\mu\text{m}^2$. This must be adjusted to give the area of non-vacuolar cytoplasm, using the ratio 133/180, equals 24·50 μm^2. The numerical density per unit area (N_A) is then $12/24\cdot50 = 0\cdot490$ mitochondria per μm^2. The volume fraction of mitochondria in this part of the cytoplasm is $13/133 = 0\cdot0977$. The shape coefficient, β, is 1·45 (from the axial ratio and Fig. 15) and the numerical density per unit volume (N_V) is

$$0\cdot490^{3/2}/1\cdot45 \ (0\cdot0977)^{1/2} = 0\cdot757 \text{ mitochondria per } \mu\text{m}^3$$

This value is not corrected for the range of mitochondria sizes present in the cell (p. 48).

$$\text{Total number of mitochondria in cell} = 0\cdot757 \times 4328\cdot1$$
$$= 3276$$
$$\text{Average mitochondrial volume is } 423\cdot0/3276 = 0\cdot129 \ \mu\text{m}^3$$

Comments

The 'average' estimates obtained are only from a single micrograph at each level. Further micrographs need to be examined in the same way and the standard errors determined to provide an indication of the accuracy of the estimates.

The nuclei in this tissue are growing and dividing, so the nuclear volumes present in the tissue will have a range of volumes with a factor of approximately two between the smallest and largest.

The plasma membrane values determined at Level I will be reasonably accurate since the walls are clearly distinguishable, but the tonoplast values from this micrograph will be subject to some uncertainty. The smallest vacuoles are not adequately resolved, while section thickness effects probably give an overestimate of the tonoplast around the larger vacuoles.

The test grid selected for the light micrograph was reasonably efficient for the point counting operation but not for the intersection counts, since the counts from the horizontal set of lines were nearly identical to those from the vertical set. The discontinuous line grid used for the electron micrograph was nearly ideal for the mitochondria, 12 profiles recording a total of 13 point counts and 13 intersection counts.

Mouse liver

Three sampling levels were used to investigate the cell structure of mouse liver hepatocytes: light micrographs at a magnification of ×1700 and electron micrographs at ×30 000 and ×58 500. The light micrographs were used for the determination of nuclear and cell volumes, while the lower magnification electron micrographs were used for estimates of mitochondrial volume and surface area. The higher magnification electron micrograph was used for area determinations of mitochondria cristae, rough and smooth endoplasmic reticulum and Golgi membranes.

Counts

Area fractions over the light micrograph (Fig. 30) were estimated using a square lattice of points. Points were counted over cytoplasm, nuclei and extracellular spaces (sinusoids and their contents).

> *Points*
> cytoplasm : 154
> nucleus : 23
> extracellular : 39
> total cell : 177
> total grid : 216

The lower magnification electron micrograph (Fig. 31, ×30 000) was analysed with a discontinuous line lattice based on equilateral rhombi. The individual line length is 1·8 cm and vertical line spacing is 1·56 cm. The number of lines present is 36, giving a total of 72 points, and a total of 21·6 μm of test lines.

Fig. 30. Mouse liver, one micron section stained with toluidine blue (×1700). Hepatocyte cells, with prominent nuclei, surround the unstained sinusoids, some of which contain occasional red blood cells. The micrograph is covered with an array of points on a 1-cm-square lattice.

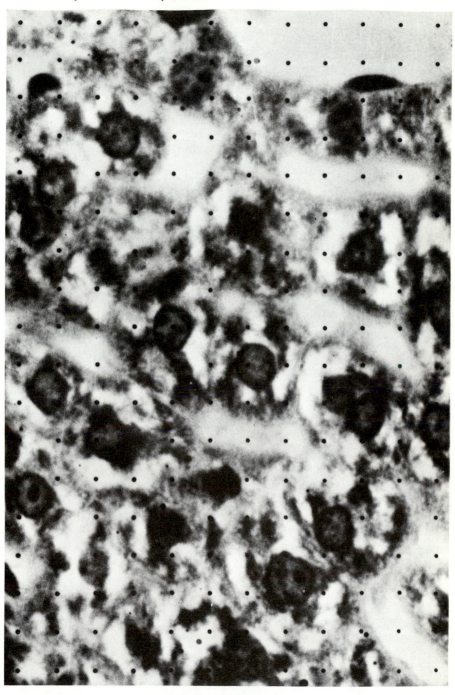

Fig. 31. Cytoplasm of mouse hepatocyte cell, with mitochondria, endoplasmic reticulum, Golgi cisternae, glycogen granules and part of a nucleus ($\times$30 000). An interrupted line grid based on a rhombus of side 1·8 cm is superimposed on the micrograph.

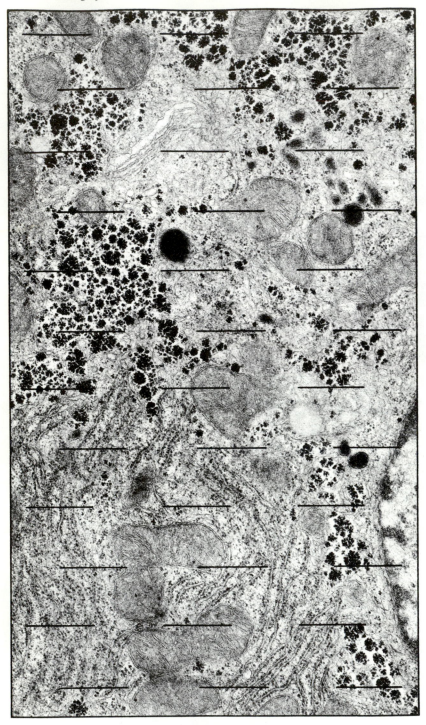

Points

mitochondria	: 12
nuclei	: 2
total cytoplasm	: 70
total grid	: 72

Intersections

mitochondrial envelope	: 14
nuclear envelope	: 2

At this magnification the endoplasmic reticulum cannot be counted accurately, nor can the Golgi membranes. The latter occur in only a single location within the cell, so that they are difficult to estimate reliably.

The higher magnification electron micrograph (Fig. 32, ×58 500) was analysed with a grid of 18 continuous lines, each 11 cm long, a total of 33·85 μm adjusted for the micrograph magnification. The micrograph was selected to include a region of the cytoplasm containing numerous mitochondria, for the estimation of mitochondrial cristae, and hence is not representative of the cytoplasm in general.

Intersections

mitochondrial envelope	: 45
cristae	: 49
endoplasmic reticulum,	
rough	: 83
smooth	: 41
Golgi membranes	: 13

Calculations: cell size, nuclear and cytoplasmic volumes

The average diameter of the nuclear profiles was found, from a series of micrographs similar to Fig. 30, to be 6·04 μm (mean of 149 measurements). The average nuclear diameter was calculated by multiplying this value by the factor $4/\pi$ to give 7·68 μm.

The Giger–Riedwyl method could not be employed for this determination as the tissue contains some tetraploid cells with double the normal nuclear volume, as well as binucleate diploid and mononucleate diploid cells.

Average radius of nuclei = 3·84 μm
Average volume of nuclei = 237·18 μm^3

This value can be used to estimate the volume of the cell, from the volume fraction of the nucleus in the cell.

Cell volume = $(177/23) \times 237·18 = 1825·9$ μm^3
Cytoplasmic volume = 1588·72 μm^3

Fig. 32. Detail of a mitochondria-rich part of a mouse hepatocyte cell (×58 000). The micrograph is overlaid with a continuous line grid with 1-cm spacing.

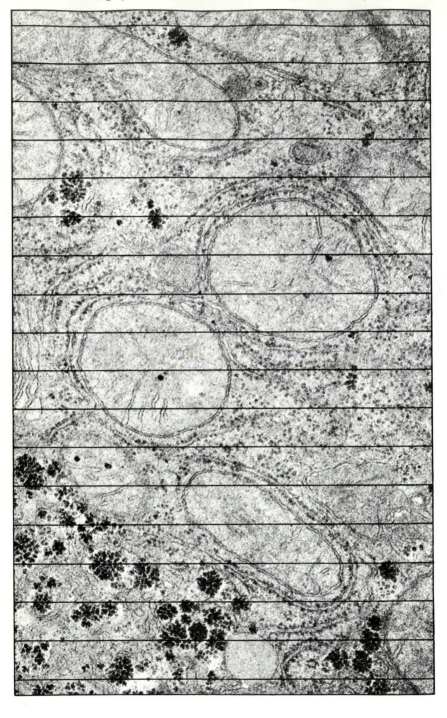

An alternative approach is to use the estimated nuclear diameter in conjunction with the numerical density of the nuclei on the micrograph to give the numerical density in the tissue.

> Number of nuclear profiles on micrograph = 20
> Area of micrograph = $100 \times 64.71 = 6471 \ \mu m^2$
> Numerical density of nuclei on micrograph = $0.0031/\mu m^2$
> Numerical density in tissue = $0.0031/7.68 = 0.004/\mu m^3$
> Volume of tissue containing one nucleus = $2500 \ \mu m^3$

Compare this value with the one determined from point counting above, adjusted to give a value for the volume of the tissue associated with a single nucleus, using the volume fraction of the nucleus in the tissue.

> Volume of tissue associated with one nucleus
> $= (216/23) \times 237.18 = 2227.0 \ \mu m^3$

These values for cell volume are lower than would be obtained from a complete analysis, due to the high area fraction of nuclear profiles on Fig. 30, which is about twice the average value.

Mitochondria

The lower magnification electron micrograph (Fig. 31) is representative of the parenchyma cell cytoplasm. The total mitochondrial volume per cell can be found from the volume fraction of the mitochondria in the cytoplasm and the cytoplasmic volume, $(12/70) \times 1588.72 = 272.3 \ \mu m^3/cell$.

Similarly the surface density of the mitochondrial envelope in the cytoplasm will give the total mitochondrial envelope area per cell. For this calculation the total line length over the cytoplasm is required. Two of the lines lie partly over the nucleus, so the line number can be taken as 35.5 and total length as $21.3 \ \mu m$.

> Surface density of mitochondria in cytoplasm
> $= (2 \times 14)/21.3 = 1.315 \ \mu m^2/\mu m^3$
> Surface area of mitochondrial envelope in cell
> $= 1.315 = 1588.72 = 2089 \ \mu m^2/cell$

At this magnification the cristae are not easily resolved. These can be resolved at a higher magnification (Fig. 32), selecting areas of cytoplasm with a large mitochondrial content to obtain a large sample of cristae. From the micrograph the ratio of mitochondrial cristae to envelope can be determined and used to find the surface density of cristae in the cytoplasm, and the total area of the cristae in the cell.

> $(49/45) \times 1.315 \times 1558.72 = 2275 \ \mu m^2$

No correction factors have been used to correct for missed membrane profiles. Mitochondrial envelope counts were recorded each time a line passed over a mitochondrion/cytoplasm boundary. Cristae membranes could have been missed in the higher magnification micrograph; a method correcting for these missed counts will be described later (p. 110).

Endoplasmic reticulum and Golgi membranes

The higher magnification electron micrograph (Fig. 32) is at about the lowest magnification that can be used for estimating the membranes of the endoplasmic reticulum and Golgi apparatus. At these magnifications only a small portion of the cytoplasm is sampled, so a larger number of micrographs are required. Even at this magnification the smooth membranes associated with the glycogen granules are not clearly defined and counts for this component are only approximate estimates.

Calculating the surface density of these components will again require the use of the total test line length over the cytoplasm, $21 \cdot 3$ μm.

Surface density of rough endoplasmic reticulum
$$= (2 \times 83)/21 \cdot 3 = 4 \cdot 90 \ \mu m^2/\mu m^3$$
Total area of rough endoplasmic reticulum $= 4 \cdot 9 \times 1588 \cdot 72$
$$= 7785 \ \mu m^2/cell$$
Surface density of smooth endoplasmic reticulum
$$= (2 \times 41)/21 \cdot 3 = 2 \cdot 42 \ \mu m^2/\mu m^3$$
Total area of smooth endoplasmic reticulum $= 2 \cdot 42 \times 1588 \cdot 72$
$$= 3845 \ \mu m^2/cell$$
Surface density of Golgi membranes $= (2 \times 13)/21 \cdot 3$
$$= 0 \cdot 768 \ \mu m^2/\mu m^3$$
Total area of Golgi membranes $= 0 \cdot 768 \times 1588 \cdot 72 = 1220 \ \mu m^2$

There have been numerous quantitative studies of liver and some of these will be reviewed later (p. 92).

Plant tapetum

The tapetum is a single layer of cells lining the inner loculus walls of the anther. In plants having the secretory type of tapetum the cells lose their vacuoles and cell walls, developing a relatively homogeneous cytoplasm packed with organelles and membrane systems that can be assessed by quantitative microscopy. At maturity, the endoplasmic reticulum is present as three morphologically distinct, but contiguous, types (Fig. 33; see Steer, 1977, for further details).

The extent of each of these endoplasmic reticulum types can be assessed

Fig. 33. (a) The endoplasmic reticulum of this tapetal cell can be divided into three morphologically distinct regions: rough cisternae (R), smooth narrow cisternae (N) and smooth distended cisternae (D). Mitochondria (M) and a plastid (P) are also present. *Avena sativa* (×42 000).

(b) Diagram of the relationships between the three forms of endoplasmic reticulum and their association with mitochondria and plastids in tapetal cells.

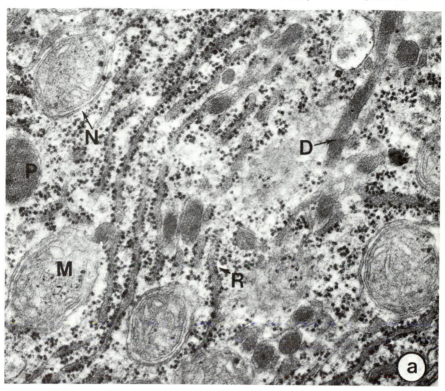

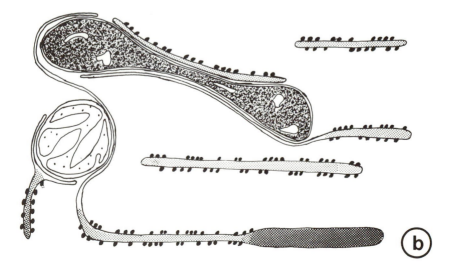

with standard line intersect grids and the surface densities determined. The individual results from 15 micrographs are presented in Table 10; the number of counts per micrograph was in the range 50–300. It can be seen from these results that the levels of each membrane type show considerable variation from one micrograph to the next. This situation is quite common and much larger ranges will often be encountered in quantitative microscopy. Means and 95% confidence limits can be calculated for each membrane system (Table 11), the calculations assume a normal distribution of the data.

Applications of quantitative microscopy

Liver
Undoubtedly the most popular organ for quantitative study has been the rat liver. Some of the earliest attempts to extract quantitative information from two-dimensional electron micrographs were carried out on this organ and these led to the development of general methods. Some of the published work on this organ will be examined in the following sections, which attempt to bring together related topics.

Table 10. *Surface densities ($\mu m^2/\mu m^3$) of endoplasmic reticulum forms in the cytoplasm of tapetal cells of* Avena sativa *(the values on each horizontal line are from a single micrograph)*

Rough	Smooth narrow	Smooth distended
4·151	1·894	0·493
5·905	1·233	1·233
4·304	1·156	0·763
3·066	0·876	0·681
2·763	1·110	0·911
7·178	1·649	0·507
6·025	1·407	0·866
4·141	1·136	0·606
4·169	1·787	0·548
3·986	1·411	0·446
5·475	1·756	0·672
8·215	1·499	0·690
5·909	0·642	0·401
3·987	1·570	0·773
6·454	2·234	0·662

Normal hepatocytes

Comprehensive analyses of normal rat liver hepatocytes were published by two groups in the late 1960s, based on methods that each had developed earlier. These will be considered in detail.

In the study by Loud (1968) three levels of sampling were used, light micrographs at a final magnification of ×1000 and electron micrographs at ×12 500 and ×50 000. A square lattice test grid was superimposed on each print using a mask in the photographic enlarger. Volume fractions were recorded using the crossover points as a square array of test points, and the surface density estimated using the two sets of lines. The micrographs were obtained from 45 tissue blocks, five blocks from each of three lobular zones from three rats. Standard deviations of the counts were determined by taking the square root of the individual count for each micrograph. Cell volumes were determined from the fraction of the parenchymal cell tissue volume occupied by nuclei, and the nuclear volume. The latter was found using a distribution curve plot of 744 nuclear profile diameters. The cell volume result obtained is actually the amount of cytoplasm per nucleus, as liver tissue contains a number of larger binucleate cells. This problem will be encountered in all the subsequent accounts on liver discussed here. Volume fractions of the major cell components were obtained routinely by counting test points. The problem of estimating the volume of innumerable small glycogen particles and rosettes was partly overcome by treating adjacent particles as part of a large contiguous whole body.

Membrane surface densities were corrected for loss of profiles, due to supposed invisibility of the membranes that pass through the sections at very oblique angles. These were allowed for by increasing membrane counts of endoplasmic reticulum by 50%. This correction was not applied to mito-chondria outer and inner membranes on the grounds that these were always counted as present whenever a test line passed over a mitochondria/cytoplasm interface, even if the membranes were indistinct. Average mitochondrial volume, and hence number per cell, was estimated by assuming that their

Table 11. *Surface densities of endoplasmic reticulum forms in the cytoplasm of tapetal cells*

Means and 95% confidence limits

Rough endoplasmic reticulum: $5.049 \pm 0.784 \ \mu m^2/\mu m^3$
Smooth narrow endoplasmic reticulum: $1.424 \pm 0.209 \ \mu m^2/\mu m^3$
Smooth distended endoplasmic reticulum: $0.683 \pm 0.108 \ \mu m^2/\mu m^3$

shape corresponded to a cylinder and using the surface to volume ratio determined from the surface density and volume fraction figures to find the diameter employing a shape coefficient.

Weibel's group (Weibel, Stäubli, Gnägi and Hess, 1969) used four levels of sampling, two at the light microscope level and two at the electron microscope level. The lowest level used direct observation of the sections at ×200, while photomicrographs at ×1000 were used at the second level. The electron micrographs used were at ×22 500 and ×90 000. Five blocks from livers of five rats were used, each block being used to provide six micrographs each for level III and IV. The two lowest levels were analysed with normal square lattice arrays of points to determine relative volume fractions of lobular and extralobular tissue and number and volume fraction of nuclei from all the cell types. Size distribution of hepatocyte nuclei were determined at the second level.

The third level used a square double lattice grid with every third line in each direction heavily ruled (ratio of points 1 : 9). The grid of 99 major points was used for volume fractions of the larger components: extracellular space, cytoplasm, nuclei and mitochondria; while the 891 fine points were used only for small scarce components, such as microbodies. These third-level micrographs were also used for surface density determinations, with a test grid of parallel continuous lines, and for determining the number of components per unit area. Mean axial ratios of mitochondrial and peroxisomal profiles were converted to ratios of intact components, and the number per unit volume of cell found using the appropriate shape coefficient (p. 48).

For the fourth level a multipurpose grid of 18 short lines was used to determine volumes and surface densities of endoplasmic reticulum and mitochondria (including cristae). These micrographs were also used to estimate bound ribosome numbers by counting their number per unit measured length of endoplasmic reticulum membrane and converting this length to an area of membrane using the section thickness.

The numerical density of nuclei, calculated using the shape coefficient was used to determine the number of hepatocytes per unit volume of liver and hence the cell volume per nucleus from the volume fraction of the hepatocytes in the whole liver.

Comparing the final results presented from these two groups shows a general agreement on the order of magnitude of the parameters measured. On the basis of an average mononuclear portion of cytoplasm ('cell') both reports come to similar conclusions about the volume of cytoplasm and the size of nuclei. Volume and numbers of mitochondria are also broadly similar. There are, however, great differences between the estimates of the surface area of endoplasmic reticulum, even though Loud's figures have been increased by a correction factor of 50%, they are still around one-third less than the uncorrected values reported by Weibel's group; these differences are discussed in a later paper from this group (Bolender *et al.*, 1978).

Hepatocytes and non-hepatocytes

While the first reports on liver concentrated on the hepatocytes, the increasing popularity of this organ for biochemical studies, in which it was often treated as a homogeneous tissue, prompted a study on the contribution of all the various cell types to the total liver population of organelles and membrane surfaces (Blouin, Bolender and Weibel, 1977). Non-hepatocytes were found to contribute a surprisingly large proportion of certain components (Table 12) to the whole liver, and hence to liver homogenates. Five animals were used in this study with a maximum of 12 blocks (84 micrographs) from each. Four sampling levels were used ($\times 7500$; $\times 14\,500$; $\times 45\,000$ and $\times 115\,000$) with square lattice test grids, counting membrane intersections with both horizontal and vertical lines. Corrections were applied for section compression (0·83) and for section thickness (p. 67).

Drug-induced changes

Cortisone. The analytical methods developed by Loud were used by his group to study the effects of cortisone, a drug known to influence liver function and structure, on hepatocyte cell structure (Wiener, Loud, Kimberg and Spiro,

Table 12. *Liver parenchyma tissue*

	Hepatocyte	Endothelial cell	Kupffer cell	Fat-storing cell
% parenchyma tissue volume				
(intercellular spaces 15·9)	77·8	2·8	2·1	1·4
% parenchyma cell volume	92·5	3·3	2·5	1·7
% aggregate membrane surface area				
Plasma membrane	73·5	15·2	4·3	7·1
Endoplasmic reticulum	93·6	3·3	2·0	1·1
Mitochondria	97·6	1·2	0·7	0·5
Golgi apparatus	84·9	11·9	2·0	1·2
Lysosomes	67·6	12·1	19·5	0·9
Pinocytotic vesicles	47·6	37·0	14·9	0·5

Composition of liver parenchyma tissue. Four cell types are present contributing varying levels of membrane to the total liver tissue (from Blouin, Bolender and Weibel, 1977).

1968). Control animals were compared with a single group exposed to cortisone for seven days. The quantitative methods were similar to those discussed above (p. 93) and revealed significant differences between control and treated animals; however, the statistical tests employed are not described. The slight increase in cell size did not give rise to any particular problems in the interpretation, for example, the areas of rough and smooth endoplasmic reticulum declined per unit volume of cytoplasm, a reduction which is not accounted for by an increase in average cell size. Mitochondria suffered pronounced morphological changes (p. 107) resulting in greater heterogeneity in the population and larger standard errors.

Phenobarbital. Weibel's group selected phenobarbital for its effects on liver structure and examined the underlying changes at the cellular level (Stäubli, Hess and Weibel, 1969) using the methods discussed earlier (p. 94). Results from control animals and those subjected to the drug for three different periods were compared using Student's two-sided t-test to compare the means of different treatments. Phenobarbital induces a marked increase in the cytoplasmic volume of the cells, 21% over five days, which is accompanied by complex changes in the levels of cell components. On a whole cell basis some remain unchanged (e.g. mitochondrial volume) while there were significant and dramatic rises in the volume and surface area of the smooth endoplasmic reticulum. Enzyme systems characteristic of the endoplasmic reticulum were analysed in parallel studies and shown to increase similarly. Comparisons of activity with membrane area were made on several bases, and the most suitable was found to be units of activity per cell compared with surface area of membranes per cell. This study emphasises the crucial importance of cell size determinations, since an increase in volume will lead to a decrease in numerical density of the component and a lower volume fraction in the cytoplasm. In the absence of cell size information these trends would be interpreted as a loss of the particular component.

Recovery from phenobarbital administration (Bolender and Weibel, 1973) leads to return of the endoplasmic reticulum levels in the cells to normal. This is associated with an increase in autophagic vacuole volume (800%) and number (96%). Livers from five animals were sampled, each providing five tissue blocks. Two sets of seven micrographs (one printed at $\times 4500$, the other at $\times 80\,000$) were taken from each block. The lower magnification prints were used for nuclear, cytoplasmic and extracellular volume fraction determination; the higher for volume fractions and surface densities of cell components. In both cases a multipurpose grid of 84 lines, 168 points, was used (actual size 729 cm^2). Section thickness effects were not routinely corrected for in the results. To minimise these problems the diameter measurements of nuclei were performed on low-magnification electron microscope prints. The resulting profile distributions were used to obtain mean nuclear diameters by either the Wicksell or Giger–Riedwyl methods (pp. 38 and 34). Comparison of the mean diameters obtained by these two methods shows a consistently lower estimate, by about 5%, using the Wicksell transformation which, they report, is a common feature of this method.

Pregnancy and lactation

Structural changes in rat liver during pregnancy and lactation have been followed by Hope (1970). One control and two test rats were used at each of five sampling times during pregnancy and six sampling times during lactation. About ten blocks from each animal were prepared and trimmed to limit the analysis to peripheral and midzonal regions, to try and reduce the variability from inclusion of central vein regions, discovered by Loud (1968). Fifty micrographs were taken for each test and control group and printed at a final magnification of $\times 12\,000$. A grid of 100 points was used to determine volume fractions. Nuclear volumes were estimated from 150 nuclear profiles in each group, measured parallel to the knife edge to avoid compression artefacts. The mean profile diameter was converted to mean sphere diameter using the factor of $4/\pi$ (p. 33). Cytoplasmic volumes were then calculated from the nuclear volume and the volume ratio of nucleus to cytoplasm determined by point counting. Changes were observed during pregnancy, especially of the rough endoplasmic reticulum, which returned to normal during lactation.

This investigation provides a useful insight into the problems of quantitative analysis. All the raw data are presented in this paper, making it an invaluable source of information. The means and standard errors are given for each group of results, but significant differences between the groups are not assessed. The limitations of using a single sampling level are revealed by examining the standard errors, which are lowest (about 5% of the mean) for intermediate size components (mitochondria) that are effectively sampled at this magnification; but high for the large-scale features (cytoplasm : nucleus ratio, 20–30%) and for the small, rarer, components (microbodies, 10–15%).

Age-related changes

Schmucker, Mooney and Jones (1977, 1978) followed the levels of endoplasmic reticulum present in rats of increasing age, using six animals at each of seven age groups. Six tissue blocks from each animal, three from the portal triad zone and three from the central vein zone, were used to provide sections for two light micrographs ($\times 1000$) and five electron micrographs at each of two magnifications ($\times 13\,800$ and $\times 45\,600$), giving a total of 2520 electron micrographs. A square lattice of points, equivalent to a total area of $10\,000\ \mu m^2$, was used to estimate volume fractions from the light micrographs, while a square double lattice of lines, with a ratio of one major to sixteen minor points, was used to estimate these parameters on the lower magnification electron micrographs. The higher magnification electron micrographs were analysed with a multipurpose test grid of short lines to determine surface densities of rough and smooth endoplasmic reticulum, and of the Golgi membranes.

Correction factors for section compression, section thickness and angle of membrane to plane of section were not employed on the grounds that these effects were minimal and, if present, applied evenly to all the data. Student's *t*-test was used to examine the significance of differences between means of membrane levels at different ages. Final analysis of the results was complicated by changes in cell volume with age, volume first increasing and then decreasing, the total membrane area per cell figures having a similar trend. On the basis of a unit volume of ground substance (cytoplasm) the rough endoplasmic reticulum remained fairly constant through life while the smooth endoplasmic reticulum showed an initial increase and then a steady decline.

Avian hepatocytes
Avian hepatocytes were examined for the effects of dantrolene sodium by Silverman and Hikida (1977). Nine control birds and four from each of six treatment durations were used in the study, selecting only a specific part of the liver in each case. Electron micrographs at $\times 13\,200$ were analysed using a plastic overlay with a square lattice test grid of 1-cm spacing for point counting. Standard errors of the means for each set of data were determined, and the means compared between experimental and control groups using analysis of variance. Significant increases in the amount of lipid were recorded, while there were losses of rough endoplasmic reticulum and mitochondria. However, these interpretations are based on volume fraction determinations, which in the absence of parallel estimates of cytoplasmic volume, cannot reliably be interpreted as reflecting changes in the whole cell. A parallel decrease in the succinic dehydrogenase activity could be taken to be a confirmation of loss of mitochondria, but this activity is also based on a section area result, using a histochemical method.

Glands
Glands contain groups of cells with specific secretory activities that can be studied using quantitative microscopy to correlate structure with secretory activity and with the effects of various stimulatory and inhibitory conditions.

The following sections examine the methods employed in some of these investigations.

Plant mucilage glands

An early study by Schnepf (1961) on *Drosophyllum* mucilage glands and a later one on *Mimulus* (Schnepf and Busch, 1976) employed similar methods to investigate the cellular dynamics underlying the secretory activity. Secretion rates were estimated by measuring the rate of increase in size of surface droplets. Whole cell measurements gave total cell volumes and total surface area of the secreting portion of the cell. Determinations, on electron micrographs, of secretory vesicle area fractions within the total cytoplasm area allowed the determination of total vesicle volume. Mean vesicle diameter measurements were used to calculate a mean volume for the vesicles, and hence their number per cell. The information was combined to give mean transit times of vesicles to the cell surface, rate of arrival at the cell surface and the mean area of new membrane delivered to the plasma membrane on fusion of the vesicles. Since the cells and plasma membrane did not increase in size it was concluded that membrane components were returning to the cytoplasm at an equivalent rate.

Pituitary

The stimulatory effect of a hypothalamic hormone, thyrotropin, on a rat pituitary gland cell line was examined in terms of the changes in cell structure accompanying prolactin synthesis and secretion (Fossum and Gautvik, 1977). Increases in volume and surface densities of endoplasmic reticulum, and in volume density of Golgi membranes, paralleled hormone synthesis and secretion. These estimates were derived from quantitative analysis of sets of electron micrographs at × 40 000. A square lattice test grid (30 mm) was used for volume fractions of mitochondria and Golgi complexes. For volume and surface densities of rough endoplasmic reticulum a multipurpose line grid on a square lattice (20-mm line length, line spacing and between rows) was used. Mean and 95% confidence levels are included in the results, but the means could not be compared by the usual statistical tests that assume a normal distribution, as the cell populations were not synchronised and contain at least two cell types. Instead the Wilcoxon rank test was employed to assess the probability of overlap between the groups of data.

Pancreas

Rat. Pancreas is a tissue that has been extensively studied structurally and biochemically, indeed it was on this tissue that the pathway of protein from endoplasmic reticulum to Golgi membranes and into secretory vesicles was established. Bolender (1974) undertook a study of this tissue to account for all cell volumes and membrane areas on either a whole cell basis or a unit volume of either cell or tissue, so that comparisons could be made with biochemical parameters or secretory activity.

An initial examination of five regions within the gland of one animal at low magnification (36 micrographs per region) did not reveal any significant differences (by an analysis of variance test) in the content of the major cell types and vessels. This meant that combining micrographs in these regions from subsequent glands would not lead to an increase in sample variability. Glands from five animals were sampled as before taking 12 micrographs from each of the five regions of each gland (see p. 63 for details of number of animals). A single section was photographed on consecutive squares of a 200 mesh grid. The negatives were printed to include the whole grid square (final magnification $\times 6200$) and the top left corner enlarged to give two prints ($\times 19\,000$ and $\times 96\,000$), a total of three sampling levels. At the first level (lowest magnification), a multipurpose test grid, 27×27cm, of 168 points (2·2-cm line length) was used for determining volume fractions. The same micrographs were used for nuclear diameter determinations, from measurements of the major and minor axes of 2300 nuclei analysed by the Giger–Riedwyl method (p. 34). The same test grid was used at the second level to estimate cytoplasmic surface densities and volume fractions by restricting the count to the areas of the grid lying over the cytoplasm. At the third level these parameters were estimated using a square double lattice test grid (1 : 9), using both sets of coarse and fine lines for the surface density count. The cell volume was determined from the nuclear volume and volume fraction of the nucleus in the cell; the tissue volume by weighing and determining its specific gravity.

The design of the analysis system enabled a complete quantitative description of the volume and surface density parameters of all the components with respect to an average cell and to the whole tissue. It should be noted that, although the data presented are not corrected for section thickness effects (p. 67), the relative amounts of different membranes can be determined with some precision and these lead to predictions about the composition of fractions derived from tissue homogenates.

Frog. The structure of the exocrine pancreas was compared in fed and fasted frogs by Slot and Geuze (1979). Feeding led to changes in the conformation, but not the surface area, of rough endoplasmic reticulum, and an increase in the volume and membrane surface area of the Golgi system and condensing vacuoles, but the number of secretory granules present in the cell was halved.

Twenty tissue blocks were taken from each of five animals in each treatment and two of these were selected at random for sectioning. The number of micrographs required at each level of sampling was determined by pooling increasing numbers of micrographs and calculating means and standard deviations at each step. This enabled selection of a final micrograph number that would give a standard error of 10% of the mean for volume fraction estimates. Test grids for volume fraction analysis consisted of a square array of points spaced 1 cm apart, and for surface density a set of parallel test lines 1 cm apart.

Three sampling levels were used, ×2400; ×12 000 and ×30 000. The lowest level was used for calculating nuclear diameters from the area within each profile (estimated using a special square grid of points, 0.2 cm apart) as the nuclear profiles from fasted frogs were irregular. A total of 500 profiles from both treatments were plotted as a histogram of 13 size classes and, using the Giger–Riedwyl method, the mean diameter and volume were determined. The latter was used in the calculation of mean cell volume, from the volume fraction of the nucleus in the cell. Although the nuclear volume increased by 25% after feeding (diameter increase from $7.28 \pm 0.77 \mu m$ to $7.87 \pm 0.75 \mu m$) the cell volume did not, greatly simplifying the interpretation of changes in the levels of cell components. The nuclear volume increase was accomplished by the re-establishment of a smooth surface, not by increasing the surface area of the envelope.

The second sampling level was used for volume fraction estimations and the highest magnification micrographs were used for surface density and surface to volume ratio determinations. At this magnification the micrographs from each animal were analysed for rough endoplasmic reticulum, secretory granules and cell membrane, while a detailed analysis of the Golgi system was achieved by photographing all Golgi profiles in randomly selected electron microscope grid squares. The surface to volume ratios of the secretory granules were used to show that their mean radius (p. 47) in each treatment was similar. The quantitative data obtained enabled detailed conclusions to be drawn about membrane flow within the Golgi system.

Parotid

Parotid gland acinar cell structure was compared before and after feeding starved rabbits, by Bedi, Cope and Williams (1974). Six animals were used for each treatment, preparing six tissue blocks per animal. Eight corners of adjacent grid squares over a single section from each block were photographed in the electron microscope and printed at ×14 700. Nuclear profile diameters were measured on light micrographs (×1750) of semi thin (0.5 μm) sections, taking the square root of the product of the major and minor axes. The resulting profile distribution was then used to obtain the mean nuclear diameter. Volume fractions of cell components, including the nucleus, were determined by point counting and the cell volume calculated from the nuclear volume and volume fraction of the nucleus in the cell. The diameters of zymogen granule profiles were measured, major and minor axes of non-circular profiles were used as above, and converted to a size–frequency distribution using the Schwartz–Saltykov method (Underwood, 1970). These were converted to numerical density by $N_V = N_A/\bar{D}$ (p. 38).

The mean volumes of the whole cell, nuclei and total mitochondria did not change on feeding. The zymogen granule volume fraction fell from 36% (450 per cell) to 22% (228 per cell). The mean diameter of the granules was slightly larger in the fed rabbits (1·18 μm v. 1·02 μm); it was concluded that some contraction in granule size occurs on maturation.

Epithelia

Mouse gallbladder epithelium

A series of studies on mouse gallbladder epithelium have been carried out by Wahlin and his associates. Changes in the cytoplasmic volume fraction occupied by mucinous secretory granules during fasting and refeeding were followed by Wahlin, Bloom, Carlsöö and Rhodin (1976). Each treatment block contained five animals, each of which was used to provide 10 randomly selected micrographs of whole cell profiles. These were analysed with a multipurpose test grid, bearing 1-cm lines, to estimate volume fractions. Standard errors of the means were determined and the observations tested for their fit to a normal distribution so that a hierarchical analysis of variance test could be applied. Similar methods were used to examine glycoprotein synthesis, and changes in the Golgi apparatus during fasting, refeeding and gallstone formation (Wahlin, 1977). Estimates of volume changes in the epithelium with diet were made using light microscope observations of 1-μm-thick sections with a square lattice grid in the eyepiece (Wahlin, 1976). Comparisons between pairs of results were made using a t-test.

Rabbit gallbladder epithelium

The comparison of normal and ouabain inhibited rabbit gallbladder epithelium is complicated by the extreme variation found in the structural organisation of this tissue and its component cells (Blom and Helander, 1977). A square lattice test grid (line spacing, corrected for micrograph magnification, 0·5–0·8 μm) was used for volume fraction determinations, and, when placed at an angle of 20° to the apical plasma membrane, for surface density estimates of the apical and lateral plasma membranes, using both sets of grid lines for intersection counts. Method errors were assessed by duplicate measurements on micrographs from one animal.

The individual epithelial cells are separated by paracellular channels of extremely variable dimensions. Their arithmetic mean width was estimated as half the mean length of lines across the channel using the sampling grid placed at random over the epithelium. Liquid flow is believed to occur in these channels and consideration is given to the correct way of averaging the channel dimensions according to the type of movement involved (bulk flow, diffusion etc.), since narrower channels will have a greater influence on decreasing flow rate than a direct proportion of their effective radius would suggest. This requires that individual values are correctly weighted before means are calculated to avoid underestimating the potential flow.

The effective increase in the surface area of the epithelium due to folding of the tissue was estimated on light micrographs by drawing a line roughly parallel to the surface of the epithelium, and using the square lattice grid to determine line intersections with this reference line and with the epithelial surface, the ratio of the two sets of counts being equal to the incremental ratio.

Human urothelium

A comparison of the organelle composition of superficial, intermediate and basal cells in human urothelium has shown distinct differences between these three cell types (Jacob, Ludgate, Forde and Selby Tulloch, 1978). Tissue from four patients was sectioned and micrographs, printed at a final magnification of $\times 25\,000$, were analysed using a square lattice of 380 points. Volume fractions were determined in each of over 30 micrographs from each cell type, as the different types occupied varying proportions of the total grid. Elements of the Golgi – endoplasmic reticulum system and polymorphic bodies showed increases from the basal to the superficial cell layers.

Muscle

Rat cardiac

Control of muscle cell structure, especially the surface to volume relationships of the whole cell and organelles, has been studied in cardiac muscle of rats by McCallister and Page (1973). They used thyroid hormone conditions that lead to an initial cessation of cell growth followed by cell enlargement. Three levels of sampling were employed ($\times 9000$; $\times 30\,000$; $\times 70\,000$), and the micrographs analysed with a square lattice grid of 1-cm spacing for the lowest and highest magnifications and 0·32cm for the remaining micrographs. Volume fractions were determined by standard point counts, and surface densities were found from intersection counts with both sets of grid lines using an equation developed by Sitte (1967).

$$S_V = (1·57/2a)(I/P)$$

where S_V is the surface density, a is the grid spacing corrected for magnification, I is the total number of intersections and P the total number of points on the grid. Mitochondrial cristae surface areas were found by the methods of Loud (described previously, p. 93; see also p. 110). The ratio of plasma membrane/unit volume of cytoplasm was found to be constant under the various growth conditions used, as was the ratio of sarcoplasmic reticulum to myofibrillar volume. Mitochondrial numbers and surface area of cristae showed marked increases under hormone treatment.

Chicken pectoralis

The problems of studying sarcoplasmic reticulum morphology in developing chicken pectoralis muscle fibres have been examined in detail by Crowe and Baskin (1977). For convenience the continuous developmental sequence was divided into five stages, though it was recognised that this increases the sampling variability. The developing system contains mixtures of highly ordered and very irregular elements, making effective sampling very difficult.

Longitudinal sections were selected to give reasonable estimates of periodic muscle structures and to reveal the organisation of the transverse tubules and sarcoplasmic reticulum. A square double lattice system, composed of lines 200 nm apart (adjusted for print magnification), with every fifth line in each direction heavily ruled, was placed on the micrographs ($\times 36\,000$) with one set of lines parallel to the long axis. Intersection counts were recorded with the horizontal and vertical set of sampling lines. Section thickness corrections were made for curvature of some of the membrane systems in the depth of the sections, especially for sarcoplasmic reticulum tubules where the overestimation was corrected by $D/D + T$ where D is the diameter of the tubule and T the section thickness. Volume changes of the tissue during fixation were recognised but not corrected for, as it was held that comparisons between muscle tissue prepared and analysed in similar ways are valid.

Means of surface density and volume fraction data were tested for variance homogeneity (Bartlett's test). Those with homogeneous variances were subjected to one-way analysis of variance tests, while those which did not have homogeneous variances were tested for correlation with age using Spearman's rank correlation test. Significant changes occur in the volume and surface density estimates for the sarcoplasmic reticulum and transverse tubules during muscle fibre development. Consideration was given to the calcium-ion-accumulating role of the sarcoplasmic reticulum, and the dependence of this process on membrane-located calcium-transporting sites. This led to the conclusion that the surface density, and total surface area are important determinants of sarcoplasmic reticulum development in skeletal muscle.

Seed development

Seeds contain a variety of storage products that can be recognised in electron micrographs, principally lipid droplets, protein bodies and starch grains in plastids (amyloplasts). These usually have regular shapes, the first two are often spherical and the third spherical or ellipsoidal. This enables calculation of numbers using the appropriate shape coefficient (p. 48). Total numbers per cell are especially significant in the case of amyloplasts since this allows determination of the timing of plastid division cycles. A series of studies has followed changes in the cell components of various seeds during development (Briarty, 1973; Briarty, Hughes and Evers, 1979; Briarty, 1980a, b).

In general several sampling levels were employed to overcome problems of scale and distribution of the cellular components in vacuolated cells. Numerical densities were determined from shape coefficients estimated from observations of intact particles, freed from the cells. Where possible direct counts were made from test grids over sample areas in the microscope, using a graticule eyepiece for the lowest, light microscope, sampling level and a test grid overlay on the screen of a Corinth microscope for the second level. This grid was a multipurpose lattice of short lines specially constructed to correct for the distortion introduced by the sloping screen of the microscope. At the third level, back projection of the original negatives, giving a final magnification on the screen of $\times 15\,000$, was used in conjunction with a double square lattice of 144 major and 576 fine points, for the intersect counts only the horizontal 'heavy' lines were used. A similar projection arrangement was used for the fourth level, with a final screen image of $\times 240\,000$ sampled by a multipurpose grid of 6×14 short lines. Total cell volumes were found from total tissue volume and an estimate of total cell number obtained by dispersing the tissue and counting the cell suspension.

Detailed information was obtained on seed development, including the relationship of division stages in plastids to cell division, and the changes occurring in endoplasmic reticulum leading up to protein body formation.

Plant transfer cells

Transfer cells possess a specialised cell wall, consisting of numerous branching, finger-like ingrowths into the peripheral cytoplasm of the cells. The resulting amplification of the plasma membrane surface area is believed to be

related to an enhanced capacity for transmembrane flow of ions and meta-
bolites (Gunning and Pate, 1974).

Estimates of the surface area amplification afforded by the wall ingrowths
were made to enable solute fluxes across the plasma membrane to be calculated from whole organ estimates of the flow involved (Gunning, Pate,
Minchin and Marks, 1974). The following example from this work is concerned with transfer cells in minor veins of leaves. These are strategically
situated alongside the sieve tubes, which are symplastic conducting channels
carrying photosynthates out of the leaves.

The standard stereological procedures were modified to cope with the
problems involved in examining parameters connected with an isolated cell
type in an organ, instead of a homogeneous tissue. The test grid area on each
micrograph ($\times 8000$) was limited to a 'cell outline', defined and drawn on the
micrograph, as the position the plasma membrane would have taken had
there been no wall ingrowths present. Points over the whole cell outline and
over the wall ingrowths alone were recorded using a test grid with a square
lattice of points. The surface density of the plasma membrane within the cell
outline was determined using a set of parallel test lines laid on each micrograph three times in different positions, recording total line length within the
cell outline in each case. The cells are approximately cylindrical in shape and
were sectioned transversely for analysis. Corrections due to this non-random
orientation are discussed, but were not employed as they only applied to a
minority of the observations. Dimensions of the whole cell were found by
using a grid of radiating lines over the micrograph with its centre at the centre
of the cell outline. Intersects with this outline were recorded and so was the
total length of the test grid within the outline. The surface density in this
system is equal to $2/r$ where r is the mean radius of the outline. For elliptical
profiles the ratio of the major and minor axes were found. The plasma membrane surface area per unit volume of cell and per unit area of cell outline was
then determined. The methods were checked by applying them to a wooden
model of known dimensions and found to be reasonably accurate.

The plasma membrane amplification factor was found to be $\times 4\cdot1$, and the
solute flux expected across this membrane was calculated as $14\cdot3$ p moles
$cm^{-2}\ s^{-1}$, slightly higher than the accepted range for plant cells. This lends
weight to the contention that the plasma membrane of cells with similar
dimensions, but lacking wall ingrowths, would be unable to support a flow of
this magnitude.

Leaf growth

In expanding leaves there is a progressive shift in emphasis from dividing
undifferentiated cells to the mature, highly differentiated condition. This

developmental sequence has been studied by following changes in the absolute surface areas of the cell membranes in expanding cucumber leaves, grown under controlled environmental conditions (Forde and Steer, 1976). Leaf area and thickness were used to determine total leaf volume at each stage. Surface densities of the membrane components were determined on sets of 100 micrographs at $\times 18\,750$, using a lattice of 480 lines. The data sets collected were found to deviate from normal distributions, so all raw data were subjected to a square root transformation. The problems generated by the formation and rapid growth of a central vacuole, causing redistribution of cell components and drastic changes in surface density levels of membranes, were considered.

Specific membrane areas were used from one of the stages of leaf expansion to calculate the expected level of membrane phospholipid in the leaf. The result differed by a factor of two from previously published total phospholipid levels in similar leaves at this stage.

Mitochondria

The structure and function of mitochondrial membranes and matrix have been the subject of intensive enquiry for some considerable time. While the matrix consists of 'soluble' enzymes, some concerned with release of respiratory carbon dioxide, the inner membranes carry the electron transport pathway, involved in the uptake of respiratory oxygen, and some other enzymes, notably succinic dehydrogenase, performing part of the carbon cycle occurring in the matrix. This level of biochemical understanding concerning mitochondrial structure and function has encouraged some cell biologists to pay particular attention to quantitating their structure, in the hope that this will provide an indicator of respiratory activity and capacity at the cellular level. For the same reasons mitochondria also make a useful model system for the study of membrane biogenesis. The following paragraphs describe the approaches and methods employed in some of these investigations.

Liver

Cortisone treatment. Changes in liver structure induced by cortisone have been discussed previously (p. 95); here a companion paper (Kimberg, Loud and Wiener, 1968) on the effects of this drug on respiratory and mitochondrial structure and function will be considered. Two or more tissue blocks from the three liver regions from each of two control and three treated rats were used to provide to total of 40 micrographs of mitochondria at a magnification of $\times 50\,000$. Making the same assumptions about mitochondrial shape as in the previous work (Loud, 1968; p. 79), the numbers, sizes and volumes of the mitochondria are fewer in number and much larger in size, this parallels a lowered respiratory activity and specific effects on a number of respiratory steps.

Thyroid hormone treatment. Another study of hormone effects on liver mitochondria involved the examination of rats under different thyroid states (Jakovcic, Swift, Gross and Rabinowitz, 1978). In general the livers of pairs of animals were taken from each treatment and used for biochemical analysis and for electron microscopy, both of intact tissue and pellets of isolated mitochondria. These were sampled at two levels, with final magnifications on the prints of ×5000 and ×50 000. Three different test grids on acetate sheets were used, two were square lattice arrays, of 5-mm and 20-mm spacing respectively, and the third was a multipurpose grid of 20-mm-long parallel lines in a staggered array on a square lattice. The test grids were clipped to 8 × 10 inch prints and numbers, volumes and surface areas of mitochondria and their membranes determined for each treatment. Numerical densities of mitochondria were determined by assuming that they were either prolate spheroids or cylindrical, and by measuring the longest length and width measurements on 10 prints from each tissue sample for 100 mitochondrial profiles from each animal. The surface area of the cristae was found using the method discussed by Yago *et al.* (1972, see p. 110). These values were compared with levels of the inner membrane markers, cardiolipin and cytochrome a, determined biochemically. It was found that increasing levels of thyroid hormone led to coupled increases in both inner membrane area and associated respiratory units.

Feeding. Changes in liver mitochondrial volume and surface to volume ratios at four time intervals after feeding rats were noted by Cieciura, Klitonczyk and Rydzynski (1978). Fifty micrographs (×81 000) were analysed from each group using either a multipurpose test lattice (line length 0·494 μm) or a square lattice of points (53·8 points/μm^2, dimensions adjusted for magnification). Point counts and line intersections with all compartments and membranes of each mitochondrion profile were recorded separately, enabling the inner and outer compartment volumes to be expressed on a unit area of inner membrane (plus cristae) basis. Differences between data groups were examined using the Kruskal–Wallis test.

Liver and heart comparisons

Normal tissue. Evidence that the area of cristae membranes is a true reflector of respiratory capacity comes from a study by Kistler and Weber (1975) in which they compared mitochondria structure and biochemistry from two different sources, the liver and heart muscle of mice. While there was a significant difference in their internal surface areas (inner membrane plus cristae), heart muscle being about twice that of liver, there was a close similarity in their ratios of membrane area to levels of succinate dehydrogenase and cytochrome aa3. The quantitative structural information was derived from four animals, half of the tissue was used as a source of isolated mitochondria, some of which were pelleted, embedded and sectioned, the remainder was fixed as intact tissue. Both preparations were sampled using four blocks from each to give 6–8 micrographs per block. A square lattice array of 144 points and a total of 624 cm of test lines was used to determine volume fractions and surface densities. Final data were expressed on a unit volume of mitochondria basis with standard errors.

Dietary and hormone controls. Two treatments designed to affect the respiratory enzyme complement and inner membrane surface area were used by Reith, Brdiczka, Nolte and Staudte (1973) to investigate the stability of membrane composition of mitochondria in rat liver and heart muscle. Riboflavin-deficient diets were used to repress flavoenzyme synthesis and triiodothyronine was used to stimulate inner membrane synthesis. Tissue samples were processed for biochemical determinations of enzyme activity and for quantitative electron microscopy.

Micrographs of thin sections taken at $\times 5000$ were enlarged and used for number and volume fraction analysis of mitochondria per cm^3 of tissue using a double lattice test grid. The numerical density was calculated using a shape coefficient ($\beta = 2.25$, p. 48) determined from measuring the 'diameter' of 100 mitochondrial profiles. Micrographs at $\times 40\,000$ were used with a multipurpose test grid for surface density estimates of mitochondrial membranes in the tissue. Correction factors for section thickness effects were not employed; this does not affect the relative levels of membrane found between different treatments.

Flavoprotein-deficient diets led to a decrease in mitochondrial membrane surface areas. When the animals were exposed to triiodothyronine the membranes increased by about 50% in surface area in the absence of parallel increases in cytochrome content. Succinic dehydrogenase levels did increase, but not at the same rate as the membrane surface area. These, and similar results for other dietary regimes, suggested that there was not a close correlation between inner membrane surface area and cytochrome content.

Adrenal cortex
A combined biochemical and structural investigation of mitochondria in regenerating rat adrenal cortex has been carried out by Yago *et al.* (1972),

comparing control animals with three stages of regeneration. Low-power micrographs (primary magnification ×40) were used to determine volume fractions of the four histological components by point counting. Whole tissue wet weight was converted to volume, and used with area fractions of parenchymal cell cytoplasm in the zona fasciculata (from planimetry of 10 electron micrographs) to estimate total volume of cytoplasm in the gland. The third level of sampling was used for collecting the diameters of over 500 mitochondrial profiles from each treatment. These were assumed to represent sections through spherical structures and the data used to determine size distribution and numerical density of the mitochondria using a method not detailed here (Bach, 1967).

The area of the cristae membranes was determined by a two-step process. The area of the outer mitochondrial membranes was unambiguously determined using a line intersect grid and scoring counts every time the line passed over the cytoplasm/mitochondrion interface, regardless of whether the membrane was visible or not. This corrected for loss of counts due to obliquely sectioned membranes (Loud, 1968; p. 79). At a higher magnification the grid was applied to a total of 50 mitochondrial profiles, scoring counts for clearly discernible outer membranes and cristae membranes. The ratio of counts obtained was multiplied by the total outer membrane surface area to give a corrected figure for the total cristae membrane area.

The amount of P-450 cytochrome in the parenchymal cells was determined using the volume fraction relationship of the cell types in the tissue. This enabled the density of the P-450 molecules on the inner membrane at all stages of regeneration to be determined. It was found that during mitochondrial proliferation their surface area increased in parallel with increasing phospholipid and protein levels to form functionally immature mitochondria. Addition of the P-450 cytochrome molecules to the pre-existing membrane at a later stage restored full mitochondrial competence, suggesting that membrane growth and differentiation is not a tightly coupled process.

Plant

Rice. Rice seedlings possess the unusual ability, for a higher plant, of being able to grow anaerobically for a period of time. This is an adaptation to seed germination in flooded conditons. Öpik (1973) has compared the respiratory activity of rice seedlings and the structure of their mitochondria under aerobic and anaerobic conditions. Respiratory activity and levels of respiratory enzymes in intact tissues were compared with the mitochondrial cross-sectional area and mitochondrial cristae density on electron micrographs of similar tissue. The areas were determined by point counting using a 0·5-cm-square grid on micrographs at ×45 000, while the number of cristae per mitochondrial profile was also recorded. Despite considerable depressions of respiratory activity and of cytochrome oxidase enzyme levels under anaerobic conditions, there was little difference in the mitochondrial size and density of cristae, indicating that there is not a close coupling between membrane synthesis and incorporation of specific enzymes.

Chlorella. John *et al.* (1973) have followed the changes in levels of cellular components during the 24-hour synchronised growth and division cycle (9 hours dark, 15 hours light) of the ellipsoid shaped unicellular alga *Chlorella*. In general this organism is a favourable material for quantitative studies since the cell sizes, and hence volumes, can be found directly from light microscope observations of intact cells, measuring the major and minor axes, while parallel biochemical studies have been carried out on many aspects of cell metabolism during the cell cycle (John *et al.*, 1973). Growth and differentiation of the inner mitochondrial membranes were studied by comparing their surface area with the levels of inner membrane enzyme activities (succinate dehydrogenase and cytochrome oxidase) at different stages of the cell cycle (Forde, Gunning and John, 1976).

Cell pellets were sectioned at each stage and the volume fractions of the mitochondria were determined using a 5-mm-square lattice on 50 prints at ×32 400; these fractions were converted to absolute volumes per cell using the whole cell volume data. Total outer membrane surface area was determined by finding the mean mitochondrial diameter of the single continuous cylindrical mitochondrion (established by three-dimensional reconstruction, Atkinson *et al.*, 1974; see p. 15) and finding the cylinder length using the volume information.

Inner to outer membrane ratios were determined from line intersect counts on 30 micrographs at each stage. A grid of three sets of parallel lines, 4·4-mm spacing with each set at an angle of 120° to the others, was used to obviate orientation effects on the count (Forde *et al.*, 1976). These counts were increased by a factor of 1·5 to correct for missing obliquely sectioned membranes. All data sets were tested for deviation from a normal distribution and those showing significant skewness were transformed before determining 95% confidence limits. The area of inner mitochondrial membrane per cell was calculated from the total outer membrane surface area and the ratio of inner to outer membrane. It was concluded that respiratory enzymes are intercalated into existing membranes, made at an earlier stage of the cell cycle, leading to a five-fold increase in the density of active enzyme molecules in the membrane.

Microsomes

Rat liver

Homogenisation of liver leads to the formation of a mixed microsome population, derived from ruptured elements of the endoplasmic reticulum and Golgi systems. Microsomes are spherical, closed membrane components, that vary in their size and surface encrustation of ribosomes, which may be fractionated by differential centrifugation.

Membrane areas in fractions. Quantitative microscopy of microsome fractions has been used to compare ribonucleic acid, protein and lipid data with surface areas of the membranes present (Wibo, Amar-Costesec, Berthet and Beaufay, 1971). The fractions were collected on filters of known diameters and the thickness of the resulting pellicle was determined microscopically allowing an estimate of the total pellicle volume to which all subsequent surface density measurements could be related. Micrographs, $\times 60\,000$, were analysed using a test grid of equidistant parallel lines, with a spacing approximately equal to the mean diameter of the profiles. The surface densities so obtained were corrected for overestimation due to section thickness effects by a factor of $1 \cdot 35$. This factor was calculated from a comparison of the value from the surface density measurement (above) with the surface area value obtained from a corrected particle profile distribution, using a particle dimension analyser and a Wicksell-type transformation (Baudhuin and Berthet, 1967). Ribosome counts were made on two bases, first on a unit area of section basis to give a number per unit volume of pellicle, using the section thickness, and second the number per microsome profile.

The results discounted a previous hypothesis that two distinct populations of microsomes, rough and smooth, overlap in the fractions to give a continuous range of ribonucleic acid levels. Instead there is a continuous spectrum of microsomal vesicles ranging from those that are entirely ribosome free to those that are maximally coated with ribosomes.

Membrane recoveries in fractions. A comparison of the level of membranes within whole rat liver to the membrane content of fractions recovered from liver homogenates was undertaken by Bolender *et al.* (1978). This involved parallel studies of intact tissue and pellicles obtained from a fractionation procedure.

The intact tissue was sampled at four levels from each of three animals. Light microscopy ($\times 200$) was used for volume fraction estimates of the tissue giving the proportion of non-parenchyma components. A multipurpose test system, 27×27 cm, containing 168 test points, was used to estimate volume fractions of nuclei and cytoplasm on 48 electron micrographs ($\times 6200$). At the third level a square double lattice system, ratio of coarse to fine points 1 : 9, was used for estimating plasma membrane surface densities on 48 micrographs ($\times 19\,000$). Intersection counts with the plasma membrane were recorded for all vertical and horizontal lines, and major points recorded over the cytoplasm. At the highest magnifications ($\times 96\,000$) surface densities of all cytoplasmic membranes were estimated on a total of 72 micrographs, probably using the same double lattice grid as in the third stage. The pellicle was analysed using the double lattice grid on micrographs at $\times 120\,000$, apart from the smallest vesicle fraction which was analysed at $\times 150\,000$.

A detailed set of correction factors was formulated and used (Weibel and Paumgartner, 1978; see p. 68) so that the absolute values of membrane area could be calculated in each case. Normally these might be expected to have similar errors, which would cancel each other out, but there are considerable changes in shape of the cell components during homogenisation, such that the pellets contain a very high proportion of spherical shapes, quite different from the flat membrane layers of the intact tissue.

The final analysis showed that most membranes could be recovered in fractions at the levels expected from the intact tissue analysis, although there was a substantial loss of mitochondrial membranes into the pool of unidentifiable smooth membranes.

Membrane identification by freeze etching. The problem of membrane identification in pellets was explored by Losa, Weibel and Bolender (1978), using the numerical density of intramembrane particles in freeze fracture preparations as an indicator of the membrane type. Intact tissue, and fractions derived from homogenates, were freeze etched and particles counted on shadow-free concave vesicle caps (P-face), using a standard test circle. This circle eliminated the smaller vesicles at the magnification used ($\times 130\,000 - 143\,000$) but had the advantage of being large enough to encompass 17 particles, giving satisfactory resolution of particle density.

In the intact tissue, endoplasmic reticulum was found to possess a characteristic particle density, midway between that of the plasma membrane and the mitochondria. By this criterion the percentage of membrane surface area attributable to endoplasmic reticulum in microsomes was found to be about 63%. This value corresponds closely to the level found from estimating the area of microsomes, identified as being of endoplasmic reticulum origin by the cytochemical marker enzyme, glucose-6-phosphatase, in thin sections by line intersect counts.

Chicken skeletal muscle

Morphological criteria were assessed and used by Scales and Sabbadini (1979) to monitor fractionation methods for heterogeneous microsome populations derived from normal and dystrophic chicken skeletal muscle. Observations were made on freeze fracture faces of intact muscle and of pelleted microsomes from different stages of the fractionation procedure. The freeze fracture faces of the vesicles are characterised by a high or low density of membrane particles on their P-faces, according to their origin from the sarcoplasmic reticulum or transverse tubule system respectively. Particles were counted within a 12-mm test circle centred over microsomal concave fracture faces on prints at × 147 000. The size of the test circle was selected to cover the range of particle densities encountered.

The membrane particle density characteristics of sarcoplasmic reticulum and transverse tubule membranes were established in intact tissue and this parameter, along with biochemical criteria, was used to assess the success of each step of a repetitive fractionation procedure designed to separate microsomes derived from these membrane systems. A final light microsome fraction was considered to be derived exclusively from the transverse tubule system, since it possessed a similar particule distribution to these membranes in intact tissue.

Conclusion

A number of important general features emerge from this review of the applications to a wide range of biological systems. Quantitative methods have been found to be useful in two types of investigation, first, those in which comparisons are made between cell structures and some biochemical or physiological process and, second, those in which comparisons are made between similar structures under different conditions.

Considerable difficulties have been experienced with the first type, comparison with biochemical or physiological processes, due to the problems of producing reliable estimates of absolute quantities of the cell components present. The main source of these problems is the estimation of appropriate correction factors to apply to the counts obtained from thin sections. The equations and graphs presented by Weibel and Paumgartner (1978) are of great value and may well be improved upon in the future. Meanwhile more attention should be paid to the approach developed by Loud (1968), whereby intersection counts with component membranes are corrected by recording line intersects with the cytoplasm/component boundary, regardless of whether or not the membranes are visible. The extension of this principle from mitochondria to other organelles should greatly improve the accuracy of surface density determinations. Most components can be distinguished from the cytoplasmic matrix by differences in staining, etc. Even endoplasmic reticulum often contains a recognisable stained internal matrix, even if ribosomes are absent from its surface. Vesicles are also distinguishable by such criteria, although the greater incidence of small cap sections may still cause difficulties.

In making comparative estimates of cell structure under different conditions the problems of correction factors are more easily resolved since the errors will be similar in each treatment and cancel out, unless gross changes occur. Such changes are the main source of complications in this type of investigation since they lead to changes in the variance levels between treatment populations, making statistical comparisons difficult.

Future progress in the field of quantitative microscopy will probably be made in two general directions, the development of improved methods of analysing micrographs and of improved methods of statistical analysis. Progress in the latter is reviewed by Nicholson (1978). Improvements in the analysis of micrographs depend on the successful application of the mathematical relationships between three-dimensional structures and two-dimensional planes (sections); work in this direction has resulted in some progress (for example, Cruz Orive, 1976; Miles and Davy, 1977, 1978; Miles, 1978; Cruz Orive and Myking, 1979) but there seems to be little doubt that the general methods developed by Loud, Weibel and others will continue to be of great value.

References

Atkinson, A.W. Jr, John, P.C.L. and Gunning, B.E.S. (1974). The growth and division of the single mitochondrion and other organelles during the cell cycle of *Chlorella*, studied by quantitative stereology and three dimensional reconstruction. *Protoplasma*, **81**, 77–109.

Bach, G. (1967). Kugelgrössenverteilung and Schnittkreisverteilung; ihre wechselseitigen Beziehungen und Verfahren zur Bestimmung der einen aus der anderen. In *Quantitative Methods in Morphology*, ed. E.R. Weibel and H. Elias, pp. 23–45. Springer, Berlin, New York.

Bailey, N.T. (1959). *Statistical Methods in Biology*. English Universities Press, London.

Baudhuin, P. and Berthet, J. (1967). Electron microscopic examination of subcellular fractions. II. Quantitative analysis of the mitochondrial population isolated from rat liver. *Journal of Cell Biology*, **35**, 631–48.

Bedi, K.S., Cope, G.H. and Williams, M.A. (1974). An electron microscopic-stereologic analysis of the zymogen granule content of the parotid glands of starved rabbits and of changes induced by feeding. *Archives of Oral Biology*, **19**, 1127–33.

Blom, H. and Helander, H.F. (1977). Quantitative electron microscopical studies on *in vitro* incubated rabbit gallbladder epithelium. *Journal of Membrane Biology*, **37**, 45–61.

Blouin, A., Bolender, R.P. and Weibel, E.R. (1977). Distribution of organelles and membranes between hepatocytes and nonhepatocytes in the rat liver parenchyma. A stereological study. *Journal of Cell Biology*, **72**, 441–55.

Bolender, R.P. (1974). Stereological analysis of the guinea pig pancreas. I. Analytical model and quantitative description of nonstimulated pancreatic exocrine cells. *Journal of Cell Biology*, **61**, 269–87.

Bolender, R.P. (1978). Correlation of morphometry and stereology with biochemical analysis of cell fractions. *International Review of Cytology*, **55**, 247–89.

Bolender, R.P., Paumgartner, D., Losa, G., Muellener, D. and Weibel, E.R. (1978). Integrated stereological and biochemical studies on hepatocytic membranes. I.

Membrane recoveries in subcellular fractions. *Journal of Cell Biology*, 77, 565–83.

Bolender, R.P. and Weibel, E.R. (1973). A morphometric study of the removal of phenobarbital-induced membranes from hepatocytes after cessation of treatment. *Journal of Cell Biology*, 56, 746–61.

Briarty, L.G. (1973). Stereology in seed development studies: some preliminary work. *Caryologia*, Supplement to 25, 289–301.

Briarty, L.G. (1975). Stereology: methods for quantitative light and electron microscopy. *Science Progress, Oxford*, 62, 1–32.

Briarty, L.G. (1980a). Stereological analysis of cotyledon cell development in *Phaseolus*. I. *Journal of Experimental Botany*, 31, 1379–86.

Briarty, L.G. (1980b). Stereological analysis of cotyledon cell development in *Phaseolus*. II. *Journal of Experimental Botany*, 31, 1387–98.

Briarty, L.G., Hughes, C.E. and Evers, A.D. (1979). The developing endosperm of wheat, a stereological analysis. *Annals of Botany*, 44, 641–58.

Calvayrac, R. and Lefort-Tran, M. (1976). Organisation spatiale des chloroplastes chez *Euglena* à l'aide de coupes sériées semi-fines. *Protoplasma*, 89, 353–58.

Cieciura, L., Klitonczyk, W. and Rydzynski, K. (1978). Stereologic analysis of mitochondria of hepatocyte from fasted rats in the course of digestion. *Folia Histochemica et Cytochemica*, 16, 193–204.

Crowe, L.M. and Baskin, R.J. (1977). Stereological analysis of developing sarcotubular membranes. *Journal of Ultrastructure Research*, 58, 10–21.

Cruz Orive, L.M. (1976). Quantifying 'pattern': a stereological approach. *Journal of Microscopy*, 107, 1–18.

Cruz Orive, L.M. and Myking, A.O. (1979). A rapid method for estimating volume ratios. *Journal of Microscopy*, 115, 127–36.

De Hoff, R.T. and Rhines, F.N. (1968). *Quantitative Microscopy*. Addison-Wesley Publishing Company, Reading, Massachusetts.

De Robertis, E.D.P., Nowinski, W.W. Saez, F.A. (1970). *Cell Biology*. Saunders Book Company, Philadelphia.

Dunn, R.F. (1972). Graphic three-dimensional representations from serial sections. *Journal of Microscopy*, 96, 301–7.

Fawcett, D.W. (1966). *An Atlas of Fine Structure. The Cell*. W.B. Saunders, London.

Forde, B.G., Gunning, B.E.S. and John, P.C.L. (1976). Synthesis of the inner mitochondrial membrane and the intercalation of respiratory enzymes during the cell cycle of *Chlorella*. *Journal of Cell Science*, 21, 329–40.

Forde, J. and Steer, M.W. (1976). The use of quantitative electron microscopy in the study of lipid composition of membranes. *Journal of Experimental Botany*, 27, 1137–41.

Fossum, S. and Gautvik, K.M. (1977). Stereological and biochemical analysis of prolactin and growth hormone secreting rat pituitary cells in culture. *Cell and Tissue Research*, 184, 169–78.

Gillis, J.-M. and Wibo, M. (1971). Accurate measurement of the thickness of ultrathin sections by interference microscopy. *Journal of Cell Biology*, **49**, 947–9.

Glauert, A.M. (1974a). *Practical Methods in Electron Microscopy. Volume 2.* North-Holland Publishing Company, Amsterdam.

Glauert, A.M. (1974b). *Practical Methods in Electron Microscopy. Volume 3.* North-Holland Publishing Company, Amsterdam.

Glauert, A.M. (1977). *Practical Methods in Electron Microscopy. Volume 6.* North-Holland Publishing Company, Amsterdam.

Glauert, A.M. (1978). *Practical Methods in Electron Microscopy. Volume 7.* North-Holland Publishing Company, Amsterdam.

Greeley, D.A. and Crapo, J.D. (1978). Practical approach to the estimation of the overall mean caliper diameter of a population of spheres and its application to data where small profiles are missed. *Journal of Microscopy*, **114**, 261–9.

Grimstone, A.V. (1977). *The Electron Microscope in Biology*, 2nd edition. Edward Arnold, London.

Gunning, B.E.S. and Hardham, A.R. (1977). Estimation of the average section thickness in ribbons of ultrathin sections. *Journal of Microscopy*, **109**, 337–40.

Gunning, B.E.S. and Pate, J.S. (1974). Transfer cells. In *Dynamic Aspects of Plant Ultrastructure*, ed. A.W. Robards, pp. 441–80. McGraw-Hill Book Company, Maidenhead, England.

Gunning, B.E.S., Pate, J.S., Minchin, F.R. and Marks, I. (1974). Quantitative aspects of transfer cell structure in relation to vein loading in leaves and solute transport in legume nodules. In *Transport at the Cellular Level*, ed. M.A. Sleigh and D.H. Jennings. *Symposium of the Society for Experimental Biology 28*, pp. 87–124. Cambridge University Press.

Gunning, B.E.S. and Steer, M.W. (1975a). *Ultrastructure and the Biology of Plant Cells.* Edward Arnold, London.

Gunning, B.E.S. and Steer, M.W. (1975b). *Plant Cell Biology: An Ultrastructural Approach.* Edward Arnold, London.

Hall, J.L. (1978). *Electron Microscopy and Cytochemistry of Plant Cells.* Elsevier/North-Holland Biomedical Press, Amsterdam.

Hall, J.L., Flowers, T.J. and Roberts, R.M. (1974). *Plant Cell Structure and Metabolism.* Longman, London.

Harris, N. (1978). Nuclear pore distribution and relation to adjacent cytoplasmic organelles in cotyledon cells of developing *Vicia faba. Planta*, **141**, 121–8.

Hayat, M.A. (1970). *Principles and Techniques of Electron Microscopy: Biological Applications, Volume 1.* Van Nostrand Reinhold Company, New York and London.

Hayat, M.A. (1972). *Principles and Techniques of Electron Microscopy: Biological Applications. Volume 2.* Van Nostrand Reinhold Company, New York and London.

Hayat, M.A. (1973). *Principles and Techniques of Electron Microscopy: Biological*

Applications. Volume 3. Van Nostrand Reinhold Company, New York and London.

Hayat, M.A. (1974). *Principles and Techniques of Electron Microscopy. Biological Applications. Volume 4*. Van Nostrand Reinhold Company, New York and London.

Hennig, A. and Elias, H. (1970). A rapid method for the visual determination of size distribution of spheres from the size distribution of their sections. *Journal of Microscopy*, **93**, 101–7.

Hope, J. (1970). Stereological analysis of the ultrastructure of liver parenchymal cells during pregnancy and lactation. *Journal of Ultrastructure Research*, **33**, 292–305.

Hopkins, C.R. (1978). *Structure and Function of Cells*. W.B. Saunders Company Limited, London.

Jacob, J., Ludgate, C.M., Forde, J. and Selby Tulloch, W. (1978). Recent observations on the ultrastructure of human urothelium. *Cell and Tissue Research*, **193**, 543–60.

Jakovcic, S., Swift, H.H., Gross, N.J. and Rabinowitz, M. (1978). Biochemical and stereological analysis of rat liver mitochondria in different thyroid states. *Journal of Cell Biology*, **77**, 887–901.

James, N.T. (1977). Stereology. In *Analytical and Quantitative Methods in Microscopy*, ed. G.A. Meek and H.Y. Elder, *Society for Experimental Biology, Seminar Series 3*, pp. 9–28. Cambridge University Press.

Jamieson, J.D., Hull, B.E., Galardy, R.E. and Maylié-Pfenninger, M.-F. (1979). Surface properties of pancreatic acinar cells: relationship to secretagogue action. In *Secretory Mechanisms*, ed. C.R. Hopkins and C.J. Duncan, *Symposia of the Society for Experimental Biology 33*, pp. 145–60. Cambridge University Press.

John, P.C.L., McCullough, W., Atkinson, A.W. Jr, Forde, B.G. and Gunning, B.E.S. (1973). The cell cycle in *Chlorella*. In *The Cell Cycle in Development and Differentiation*, ed. M. Balls and F.S. Billet, *Symposium of the British Society for Developmental Biology 1*, pp. 61–76. Cambridge University Press.

Jordan, E.G. and Saunders, A.M. (1976). The presentation of three dimensional reconstructions from serial sections. *Journal of Microscopy*, **107**, 205–6.

Kimberg, D.V., Loud, A.V. and Wiener, J. (1968). Cortisone-induced alterations in mitochondrial function and structure. *Journal of Cell Biology*, **37**, 63–79.

Kistler, A. and Weber, R. (1975). A morphometric analysis of inner membranes related to biochemical characteristics of mitochondria from heart muscle and liver in mice. *Experimental Cell Research*, **91**, 326–32.

Ledbetter, M.C. and Porter, K.R. (1970). *Introduction to the Fine Structure of Plant Cells*. Springer-Verlag, Berlin, Heidelberg, New York.

Losa, G.A., Weibel, E.R. and Bolender, R.P. (1978). Integrated stereological and biochemical studies on hepatocytic membranes. III. Relative surface of endoplasmic reticulum membranes in microsomal fractions estimated on freeze-fracture preparations. *Journal of Cell Biology*, **78**, 289–308.

Loud, A.V. (1968). A quantitative stereological description of the ultrastructure of normal rat liver parenchymal cells. *Journal of Cell Biology*, **37**, 27–46.

Loud, A.V., Barany, W.C. and Pack, B.A. (1965). Quantitative evaluation of cytoplasmic structures in electron micrographs. *Laboratory Investigation*, **14**, 996–1008.

McCallister, L.P. and Page, E. (1973). Effects of thyroxin on ultrastructure of rat myocardial cells: a stereological study. *Journal of Ultrastructure Research*, **42**, 136–55.

Miles, R.E. (1978). The sampling, by quadrats, of planar aggregates. *Journal of Microscopy*, **113**, 257–67.

Miles, R.E. and Davy, P. (1977). On the choice of quadrats in stereology. *Journal of Microscopy*, **110**, 27–44.

Miles, R.E. and Davy, P. (1978). Particle number or density can be stereologically estimated by wedge sections. *Journal of Microscopy*, **113**, 45–51.

Nicholson, W.L. (1978). Application of statistical methods in quantitative microscopy. *Journal of Microscopy*, **113**, 223–39.

Novikoff, A.B. and Holtzman, E. (1976). *Cells and Organelles*, 2nd edition. Holt, Rinehart & Winston, New York.

Öpik, H. (1973). Effect of anaerobiosis on respiratory rate, cytochrome oxidase activity and mitochondrial structures in coleoptiles of rice (*Oryza sativa* L.). *Journal of Cell Science*, **12**, 725–39.

Parker, R.E. (1979). *Introductory Statistics for Biology*, 2nd edition. Edward Arnold, London.

Pellegrini, M. (1976). Présence d'une mitochondrie unique et de protoplastes isolés chez l'*Euglena gracilis* Z, en culture synchrone hétérotrophe, à l'obscurité. *Comptes Rendues de l'Academie des Sciences, Paris, Série D*, **283**, 911–13.

Pellegrini, M. and Pellegrini, L. (1976). Continuité mitochondriale et discontinuité plastidale chez l'*Euglena gracilis* Z. *Comptes Rendues de l'Academie des Sciences, Paris. Série D*, **282**, 357–60.

Pickett-Heaps, J.D. (1975). *Green Algae: Structure, Reproduction and Evolution in Selected Genera*. Sinauer Associates Incorporated, Sunderland, Massachusetts.

Porter, K.R. and Bonneville, M.A. (1973). *Fine Structure of Cells and Tissues*, 4th edition. Lea and Febiger, Philadelphia.

Poux, N., Favard, P. and Carasso, N. (1974). Étude en microscopie électronique haute tension de l'appareil vacuolaire dans les cellules méristématiques de racines de concombre. *Journal de Microscopie*, **21**, 173–80.

Rakic, P., Stensas, L.J., Sayre, E.P. and Sidman, R.L. (1974). Computer-aided three-dimensional reconstruction and quantitative analysis of cells from serial electron microscopic montages of foetal monkey brain. *Nature, London*, **250**, 31–4.

Reid, N. (1974). Ultramicrotomy. In *Practical Methods in Electron Microscopy, Volume 3*, ed. A.M. Glauert, pp. 211–353. North-Holland Publishing Company, Amsterdam.

Reith, A., Brdiczka, D., Nolte, J. and Staudte, H.W. (1973). The inner membrane of mitochondria under influence of triiodothyronine and riboflavin deficiency in rat heart muscle and liver. *Experimental Cell Research*, **77**, 1–14.

Rohr, H., Oberholzer, M., Bartsch, G. and Keller, M. (1976). Morphometry in experimental pathology: methods, baseline data, and applications. *International Review of Experimental Pathology*, **15**, 233–325.

Rose, P.E. (1980). Improved tables for the evaluation of sphere size distributions including the effect of section thickness. *Journal of Microscopy*, **118**, 135–41.

Sandler, S.S. (1974). Direct three-dimensional reconstruction of a corneal stromal lamella from electron micrographs. *Journal of Theoretical Biology*, **48**, 207–13.

Scales, D.J. and Sabbadini, R.A. (1979). Microsomal T system. A stereological analysis of purified microsomes derived from normal and dystrophic skeletal muscle. *Journal of Cell Biology*, **83**, 33–46.

Schmucker, D.L., Mooney, J.S. and Jones, A.L. (1977). Age-related changes in the hepatic endoplasmic reticulum: a quantitative analysis. *Science*, **197**, 1005–7.

Schmucker, D.L., Mooney, J.S. and Jones, A.L. (1978). Stereological analysis in the Fischer – 344 rat. Influence of sublobular location and animal age. *Journal of Cell Biology* **78**, 319–37.

Schnepf, E. (1961). Quantitative Zusammenhänge zwischen der Sekretion des Fangschleines und den Golgi-Strukturen bei *Drosophyllum lusitanicum*. *Zeitschrift für Naturforschung*, **16b**, 605–10.

Schnepf, E. and Busch, J. (1976). Morphology and kinetics of slime secretion in glands of *Mimulus tilingii*. *Zeitschrift für Pflanzenphysiologie*, **79**, 62–71.

Schotz, F. (1972). Dreidimensionale, mass stabgetreue. Rekonstruktion einer grünen Flagellatenzelle nach Elektronenmikroskopie. *Planta*, **102**, 152–9.

Silverman, H. and Hikida, R.S. (1977). Membrane systems of avian hepatocytes during chronic exposure to dantrolene sodium: a morphometric, ultrastructural and histochemical study. *Tissue and Cell*, **9**, 507–20.

Sitte, H. (1967). Morphometrische Untersuchungen an Zellen. In *Quantitative Methods in Morphology*, ed. E.R. Weibel and H. Elias, pp. 167–98. Springer, Berlin, New York.

Slot, J.W. and Geuze, J.J. (1979). A morphometrical study of the exocrine pancreatic cell in fasted and fed frogs. *Journal of Cell Biology*, **80**, 692–707.

Small, J.V. (1968). Measurement of section thickness. In *Proceedings of the 4th European Congress on Electron Microscopy, Volume 1*, pp. 609–10. Tipografia Poliglotta Vaticana, Rome.

Snedecor, G.W. and Cochran, W.G. (1967). *Statistical Methods*, 6th edition. Iowa State University Press.

Sokal, R.R. and Rohlf, F.J. (1969). *Biometry*. W.H. Freeman, San Francisco.

Solari, A. (1973). Étude quantitative d'organes ou de tissus. I. Méthodes d'estimations des volumes. *Annales de Biologie Animale, Biochimie, Biophysique*, **13**, 247–65.

Solari, A. (1975). Étude quantitative d'organes ou de tissus. II. Méthodes de mesure des surfaces de structures cellulaires. *Annales de Biologie Animale, Biochimie, Biophysique*, **15**, 589–605.

Stäubli, W., Hess, R. and Weibel, E.R. (1969). Correlated morphometric and biochemical studies on the liver cell. II. Effects of phenobarbital on rat hepatocytes. *Journal of Cell Biology*, **42**, 92–112.

Steer, M.W. (1977). Differentiation of the tapetum in *Avena*. II. The endoplasmic reticulum and golgi apparatus. *Journal of Cell Science*, **28**, 71–86.

Stevens, B.J. (1977). Variation in number and volume of the mitochondria in yeast according to growth conditions. A study based on serial sectioning and computer graphics reconstructions. *Biologie Cellulaire*, **28**, 37–56.

Thiéry, G. (1979). Colorations signalétiques électives sur coupes épaisses du réticulum endoplasmique, de la chromatine et des surfaces cellulaires libres des cellules animales. *Biologie Cellulaire*, **35**, 159–64.

Thiéry, G. and Bergeron, M. (1977). Études sur coupes épaisses des polysaccharides acides de la cellule du tube contourné proximal du rein de rat. *Biologie Cellulaire*, **30**, 279–82.

Threadgold, L.T. (1976). *The Ultrastructure of the Animal Cell*, 2nd edition. Pergamon Press, Oxford.

Underwood, E.E. (1968). Particle size distribution. In *Quantitative Microscopy*, ed. R.T. De Hoff and F.N. Rhines, pp. 149–200. McGraw-Hill Book Company, London.

Underwood, E.E. (1970). *Quantitative Stereology*. Addison-Wesley Publishing Company, Reading, Massachusetts.

Wahlin, T. (1976). Effects of lithogenic diets on mouse gallbladder epithelium. *Virchows Archives B, Cell Pathology*, **22**, 273–86.

Wahlin, T. (1977). Synthesis of glycoproteins in the Golgi complex of the mouse gallbladder epithelium during fasting, refeeding, and gallstone formation. *Histochemistry*, **51**, 133–40.

Wahlin, T., Bloom, G.D., Carlsöö, B. and Rhodin, L. (1976). Effects of fasting and refeeding on secretory granules of the mouse gallbladder epithelium. *Gastroenterology*, **70**, 353–8.

Ware, R.W. and Lopresti, V. (1975). Three dimensional reconstruction from serial sections. *International Reviews of Cytology*, **40**, 325–440.

Weibel, E.R. (1963). *Morphometry of the Human Lung*. Springer, Berlin.

Weibel, E.R. (1969). Stereological principles for morphometry in electron microscopic cytology. *International Reviews of Cytology*, **26**, 235–302.

Weibel, E.R. (1974). Stereological quantitation of cellular membrane systems. In *Electron Microscopy 1974, Volume II*, ed. J.V. Sanders and D.J. Goodchild, pp. 12–13. The Australian Academy of Science, Canberra.

Weibel, E.R. (1979). *Stereological Methods. Volume 1. Practical Methods for Biological Morphometry*. Academic Press, London.

Weibel, E.R. and Bolender, R.P. (1973). Stereological techniques for electron microscopic morphometry. In *Principles and Techniques of Electron Microscopy*.

Biological Applications, Volume 3, ed. M.A. Hayat, pp. 237–96. Van Nostrand Reinhold Company, New York.

Weibel, E.R. and Paumgartner, D. (1978). Integrated stereological and biochemical studies on hepatocytic membranes. II. Correction of section thickness effect on volume and surface density estimates. *Journal of Cell Biology*, **77**, 584–97.

Weibel, E.R., Stäubli, W., Gnägi, H.R. and Hess, F.A. (1969). Correlated morphometric and biochemical studies on the liver cell. I. Morphometric model, stereologic methods, and normal morphometric data for rat liver. *Journal of Cell Biology*, **42**, 68–91.

Wibo, M., Amar-Costesec, A., Berthet, J. and Beaufay, H. (1971). Electron microscope examination of subcellular fractions. III. Quantitative analysis of the microsomal fraction isolated from rat liver. *Journal of Cell Biology*, **51**, 52–71.

Wicksell, S.D. (1925). On the size distribution of sections of a mixture of spheres. *Biometrika*, **17**, 84–99.

Wiener, J., Loud, A.V., Kimberg, D.V. and Spiro, D. (1968). A quantitative description of cortisone-induced alterations in the ultrastructure of rat liver parenchymal cells. *Journal of Cell Biology*, **37**, 47–61.

Williams, M.A. (1977). Quantitative methods in biology. In *Practical Methods in Electron Microscopy, Volume 6*, ed. A.M. Glauert, part II. North-Holland Publishing Company, Amsterdam.

Wolfe, S.L. (1972). *Biology of the Cell*. Wadsworth Publishing Company, Belmont, California.

Yago, N., Seki, M. Sekiyama, S., Kobayashi, S., Kurokawa, H., Iwai, Y., Sato, F. and Shiragai, A. (1972). Growth and differentiation of mitochondria in the regenerating rat adrenal cortex. A correlated biochemical and stereological approach. *Journal of Cell Biology*, **52**, 503–13.

Index

actin, 31
adrenal cortex, 109
amyloplasts, 105
autophagic vacuoles, 96
avian hepatocytes, 98

cardiac muscle, 103
cardiolipin, 108
cell cycle, 111
cell wall, 11, 57
cells; evolution of, 4; structure of, 3, 5;
    components of, 11; volume of, 65
Chlorella, 14, 35, 111
chloroplasts, 24
collagen, 31
compartmentation, 4
confidence limits, 60
contrast, 5
correction factors, 66, 115
cortisone, 95, 107
counting, 57
cubes, 24, 48
cytochrome, 108–10
cytochrome oxidase, 111
cytoplasmic volume, 65, 81, 87

dehydration, 66
dimensions, linear, 24
dynamic activities, 18

electron microscopes, 6, 14, 17; high voltage,
    14
ellipse, 24
embedding, 6

endoplasmic reticulum, 12, 24, 51, 55, 68, 90,
    105, 112, 113; *see also* sarcoplasmic
    reticulum
enzymes, 4
epithelium, 102
experimental design, 48

fixation, 6, 66
flavoenzyme, 109
fluorescence, 6
freeze fracture, 113
frog, 100

gallbladder, 102
genetic code, 4
glands, 98
glucose-6-phosphatase, 113
glycogen, 11, 86
glycoprotein, 102
Golgi cisternae, 12, 35, 67, 90, 112
Golgi vesicles, 12, 16, 17, 40, 41, 47

heart, 108

image formation, 5
image interpretation, 7

lactation, 97
leaves, 106
length, estimation of, 31
light microscopy, 5
line intersect grids, 55, 78, 81, 87, 109, 111,
    112
liver, 65, 84, 92, 107